3

Teacher's Book

Karen Morrison
Lisa Greenstein

Great Clarendon Street, Oxford, OX2 6DP, United Kingdom

Oxford University Press is a department of the University of Oxford.

It furthers the University's objective of excellence in research, scholarship, and education by publishing worldwide. Oxford is a registered trade mark of Oxford University Press in the UK and in certain other countries.

British Library Cataloguing in Publication Data

Data available

ISBN: 978-1-382-01014-6

1 3 5 7 9 10 8 6 4 2

Paper used in the production of this book is a natural, recyclable product made from wood grown in sustainable forests. The manufacturing process conforms to the environmental regulations of the country of origin.

Printed in Great Britain by CPI

Acknowledgements

The publisher and authors would like to thank the following for permission to use photographs and other copyright material:

Cover: Matthieu Nivesse. **Photos: p20:** Oxford University Press.

Artwork by Joseph Wilkins, John Haslam, Q2A Media, Integra Software Services, Pantek Media, and Oxford University Press.

Every effort has been made to contact copyright holders of material reproduced in this book. Any omissions will be rectified in subsequent printings if notice is given to the publisher.

Cover activities

The following activities are based on the Level 3 Pupil Book and Workbook cover image. You can use these stimulus questions according to the children's learning to date.

Geometry

Spotting shapes, lines and angles: How many different 2D and 3D shapes can the children see? Can they identify 2D faces on the 3D shapes? Encourage the children to describe the properties of the shapes – ask how many faces, edges and vertices they have. Can the children identify right angles and angles bigger/smaller than right angles? Talk about the amount of turn the cranes might make. Look at the tessellating triangle patterns. Ask what amount of turn the triangles have moved through. How many parallel, perpendicular, horizontal and vertical lines can the children find?

Recognising patterns: Ask questions such as: *Can you see any repeating patterns? What is the rule? What would come next if you extended the pattern?*
Encourage the children to notice patterns where the objects change shape, size or orientation, not just colour – for example, the tessellating triangles on the buildings or the growing patterns made by the vehicles and the posts next to the construction site.

Measure

Roman numerals and time: Ask questions such as: *Are there any numbers in the picture? Which numbes can you see? What time does the clock show? What time will it be one hour later?*

Contents

How Nelson Maths works

LEVEL	PUPIL BOOK	WORKBOOKS	TEACHING SUPPORT	DIGITAL CONTENT ON OXFORD OWL
STARTER LEVEL				For all levels: • Digital versions of the Pupil Books, Workbooks and Teacher's Books • Assessment support • Parent notes • Curriculum mapping and planning guides • Vocabulary support
1				
2				
3				
4				
5				
6				

Access the digital content for this course online at
www.oxfordowl.co.uk

How to use Nelson Maths

Nelson Maths is a comprehensive maths programme for children aged 4–11. The course covers the following five strands: Number, Measure, Geometry, Statistics and Algebra to ensure full coverage of your chosen curriculum. The course is also full of opportunities to develop children's problem-solving skills through carefully designed questions and activities.

Nelson Maths is made up of seven levels (one for each year group) and Level 3 contains 18 units, which should ideally be taught in order. The units have been arranged in a careful progression, designed to support children to build their skills and knowledge and make connections across the different strands. Your children may progress through some units more quickly than others, but it is recommended that you spend approximately two weeks on each unit.

Children each have a **Pupil Book** and write-in **Workbook**. The main lesson activities are included in the **Pupil Book**, and the **Workbook** includes extra practice and consolidation activities. The **Teacher's Book** contains detailed teaching support for each unit, which will help you to introduce new concepts, revise prior learning, deepen children's understanding of a topic and encourage mathematical conversations and discussion.

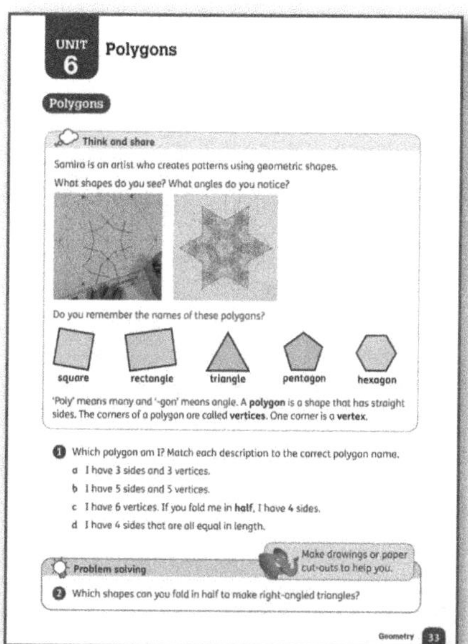

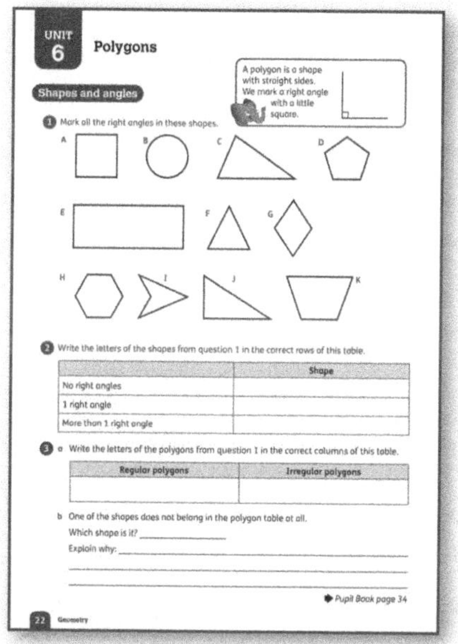

Each unit in the **Pupil Book** is supported by a corresponding unit in the **Workbook**.

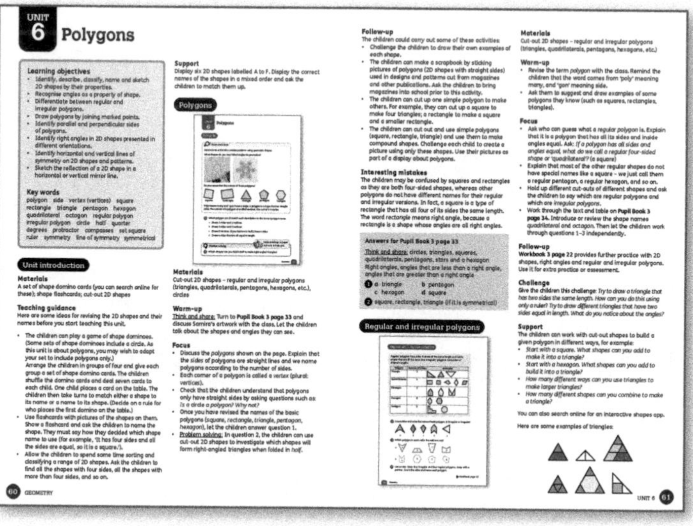

The **Teacher's Book** provides detailed teaching guidance for each unit and answers to the questions and activities.

Read the relevant unit of the Teacher's Book before you begin teaching it. It's important to familiarise yourself with the learning objectives, key vocabulary and resources you need before each lesson. The following pages will give you a good understanding of how to use the Pupil Book, Workbook and Teacher's Book together.

In addition to the printed materials, you can also find lots of supplementary digital resources online on Oxford Owl. Please see page 9 for more information.

Using the Pupil Books

The **Pupil Book** units are designed to be worked through in order, following the discussion prompts and activities suggested in the **Teacher's Book**, and using the linked **Workbook** pages to help children to consolidate their understanding through independent practice. The following features of the **Pupil Books** are designed to help you get the most out of your lessons.

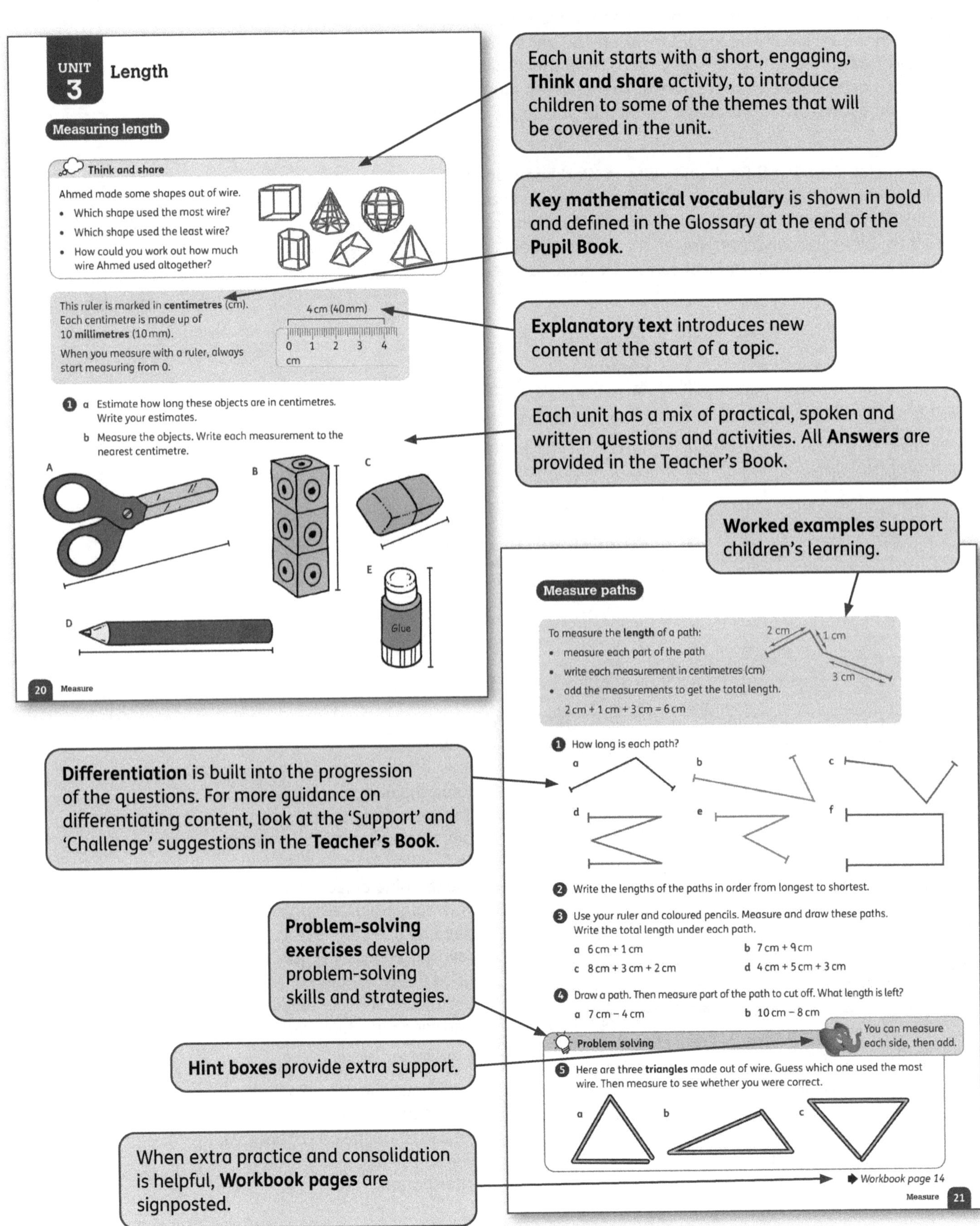

Each unit starts with a short, engaging, **Think and share** activity, to introduce children to some of the themes that will be covered in the unit.

Key mathematical vocabulary is shown in bold and defined in the Glossary at the end of the **Pupil Book**.

Explanatory text introduces new content at the start of a topic.

Each unit has a mix of practical, spoken and written questions and activities. All **Answers** are provided in the Teacher's Book.

Worked examples support children's learning.

Differentiation is built into the progression of the questions. For more guidance on differentiating content, look at the 'Support' and 'Challenge' suggestions in the **Teacher's Book**.

Problem-solving exercises develop problem-solving skills and strategies.

Hint boxes provide extra support.

When extra practice and consolidation is helpful, **Workbook pages** are signposted.

The first unit in each Level is called **Think maths**. It encourages a growth mindset and builds resilience by teaching children that mistakes are a positive part of their learning journey. It prepares children to think mathematically and make connections across maths.

Each book has three **Mixed practice** reviews – one per 'term'. These formative assessment questions help you to assess children's understanding. You may choose to revisit some concepts after the practice.

Mixed practice **Answers** are provided in the **Teacher's Book**.

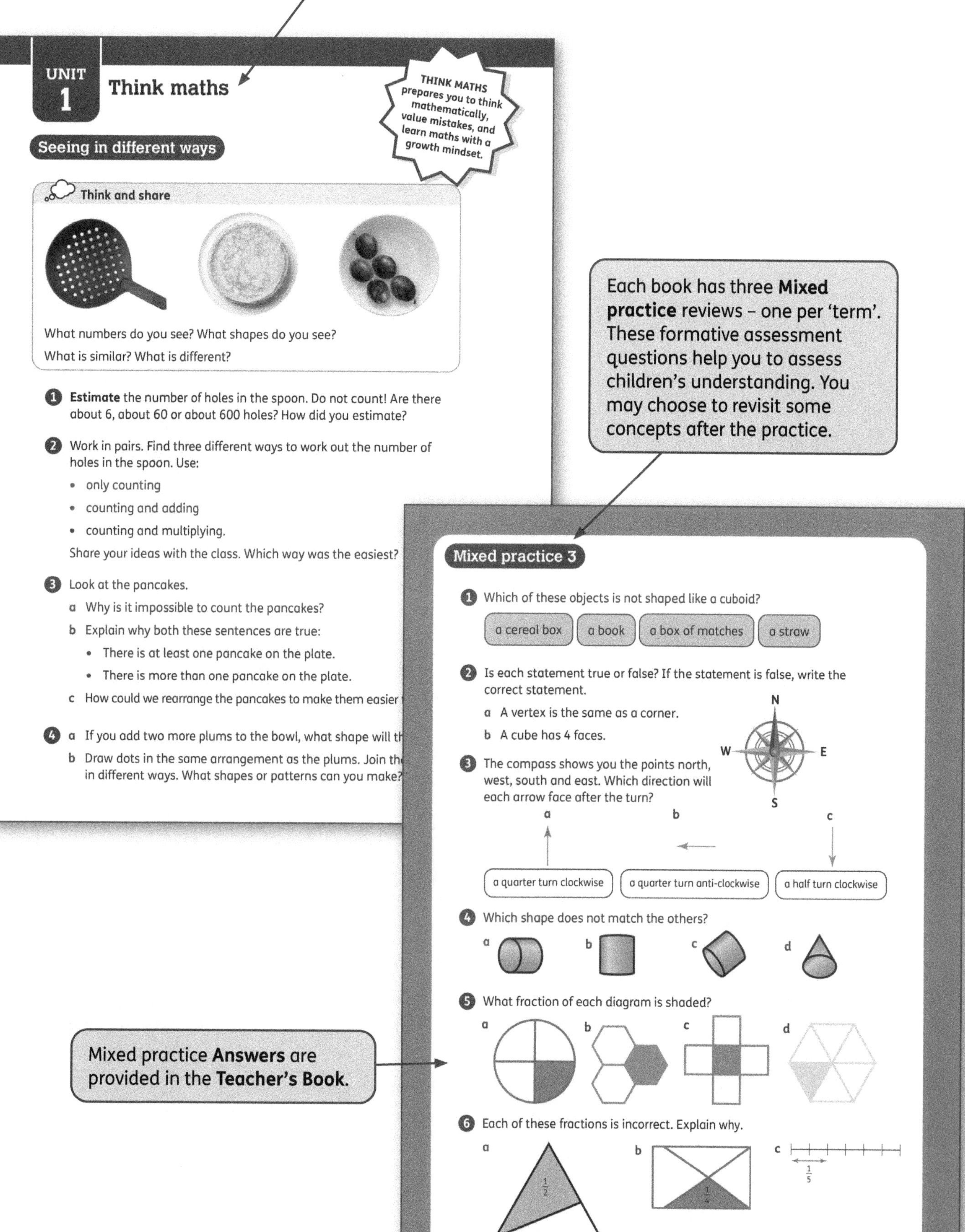

Using the Workbooks

The **Workbooks** provide essential practice to help children consolidate their learning. They provide new questions and activities linked to the concepts introduced in the **Pupil Books**. They are designed for independent use, making them ideal for homework or additional classwork. We recommend that children complete each page in the **Workbook** after they have completed the corresponding lesson in the **Pupil Book**. When extra practice and consolidation is helpful, **Workbook** pages are signposted at the end of the Pupil Book lesson.

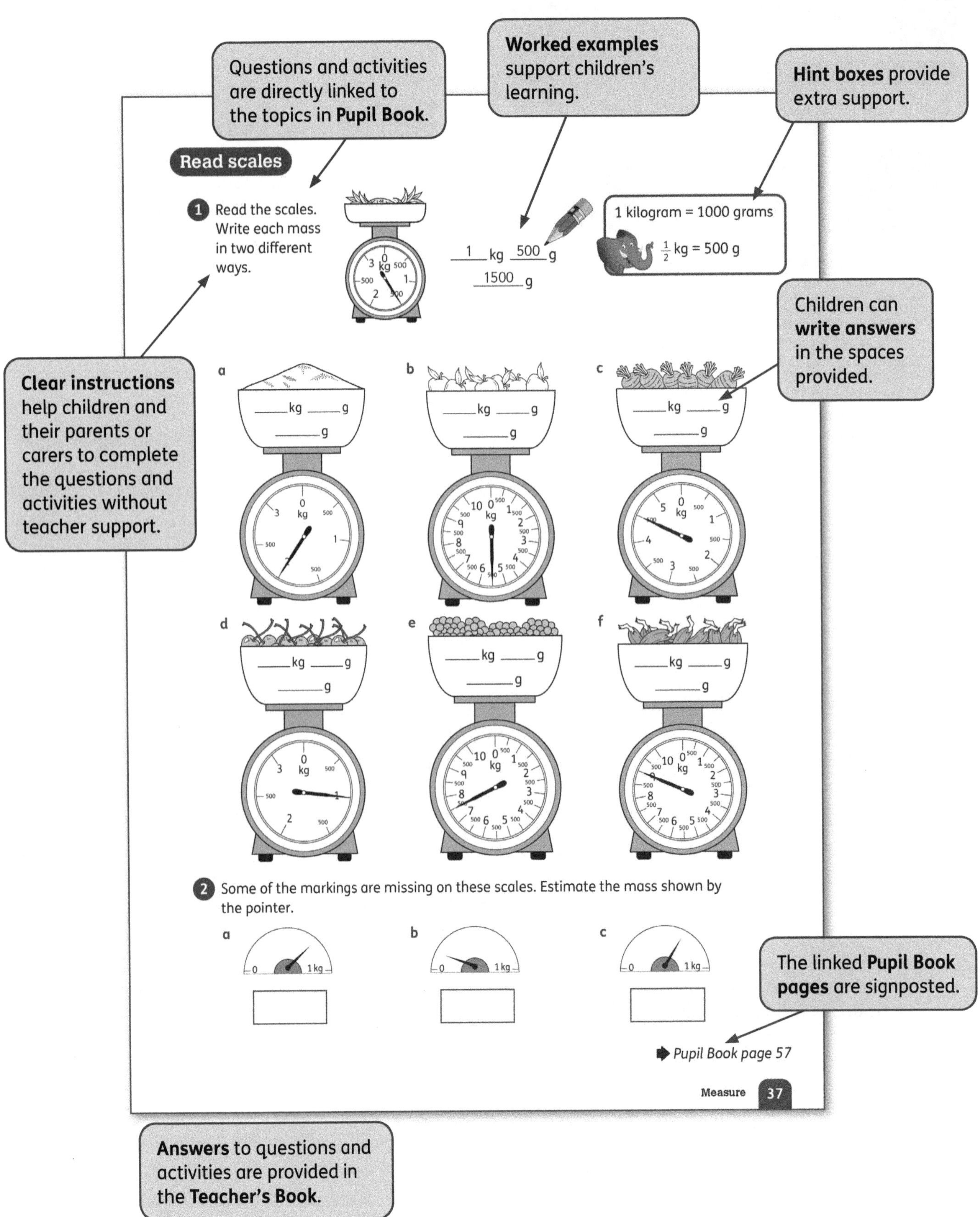

Using the Digital Content

This edition of *Nelson Maths* is supported by additional digital resources available online on **Oxford Owl**. These include:

- Digital versions of the Pupil Books, Workbooks and Teacher's Books to support planning and for front of class display
- A set of printable assessments with a progress tracking tool and mark scheme
- Guidance for you to share with children's parents or carers about their child's learning, including support for homework and ideas for incorporating maths into everyday life
- Vocabulary support
- Curriculum correlation charts
- Planning support

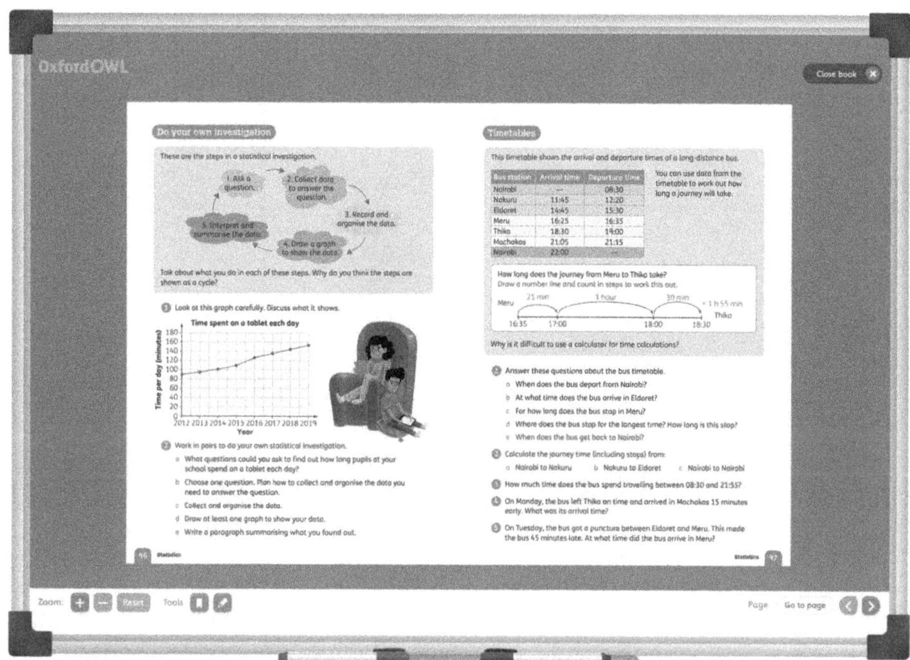

Using the Teacher's Book

The **Teacher's Book** provides step-by-step lesson notes for each unit in the **Pupil Book** and **Workbook**, including learning objectives, answers to the questions activities and suggestions for additional challenge and support.

Before you start teaching the units, take some time to familiarise yourself with these useful sections at the start of the **Teacher's Book**. These four sections are designed to help you get the most out of the *Nelson Maths* teaching materials:

1. **Introduction** – find out more about the mathematical strands, skills and fundamental principles that underpin *Nelson Maths*. Get ideas and support for teaching maths in an inclusive way that acknowledges individual differences and human diversity.

2. **Teaching approach** – guidance for teaching *Nelson Maths* effectively, including:
- how to develop robust learners with a growth mindset and an understanding of the importance of making mistakes
- the importance of the do-talk-record model that underpins the lessons
- using engaging teaching methods, from investigations and number talks, to exploring maths in real life contexts.

3. **An environment for exploration and play** – practical support on organising your classroom space; approaching whole-class, group and individual activities; and using manipulatives and games to deepen children's learning.

4. **Activity bank of warm-ups and mental maths** – a bank of engaging mental maths, support, and consolidation activities to use with your class.

Teacher's Book: using the unit pages

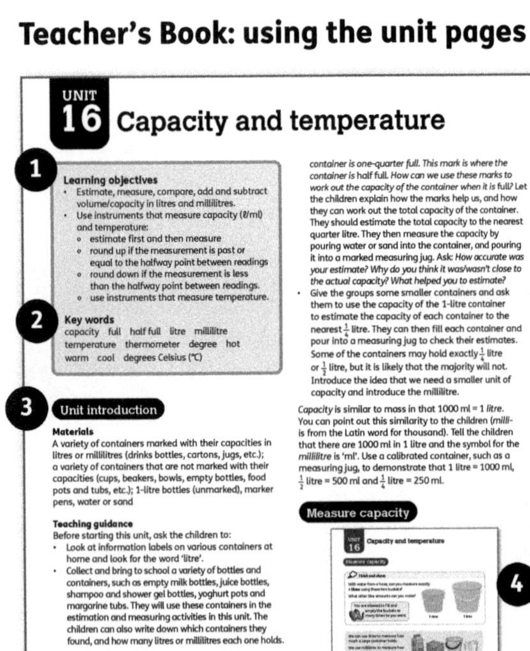

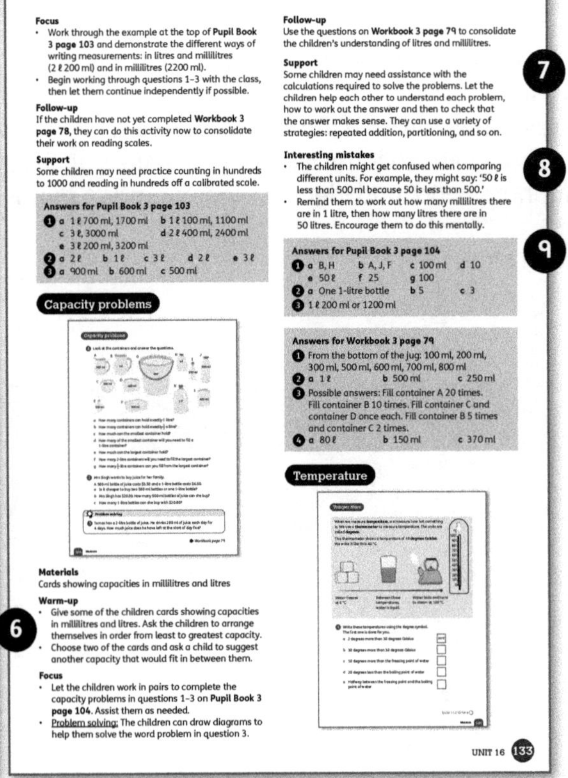

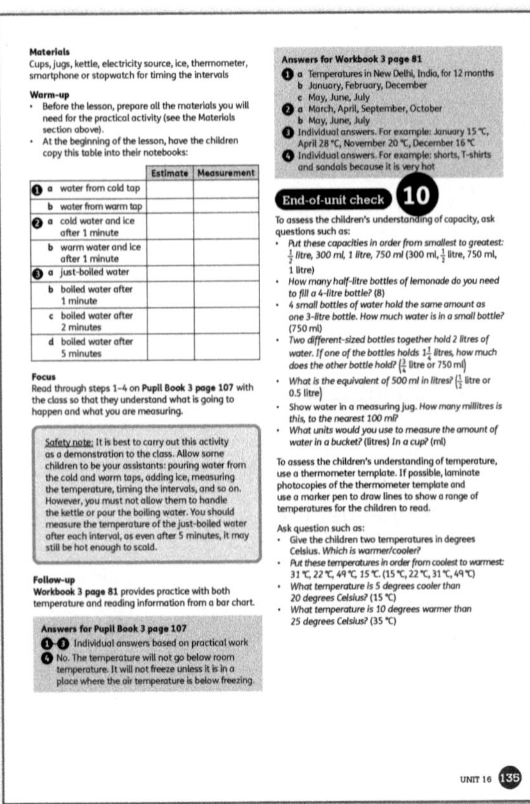

1 **Learning objectives** are listed at the beginning of each unit.

2 **Key words** match the important words highlighted in bold in the **Pupil Book**.

3 **Unit introduction** activities introduce the topics in the unit, engage children and find out what they already know.

4 The **linked Pupil Book and Workbook pages** are signposted.

5 **Materials** needed for the lesson are listed.

6 Most lessons are broken down into three stages:

- **Warm-up** – introduce the topic and unlock prior learning
- **Focus** – develop a firm understanding of concepts and skills by working through the **Pupil Book**
- **Follow-up** – recap the key learning points and extend learning with the activity suggestions in the **Workbook**.

7 Where relevant, **Challenge** and **Support** sections give guidance on deepening the learning or making it more accessible, so that every child is catered for.

8 It is important for children to see mistakes as part of their learning journey. The **Interesting mistakes** section highlights common errors and misconceptions that children may have, along with what to look for and advice on how to support children.

9 **Answers** for **Pupil Book** and **Workbook** questions are provided to make marking easy.

10 An **End-of-unit check** allows you to assess how well children have understood the concepts in each unit, so you can monitor children's understanding throughout the course. A set of formative assessment questions or a closing activity are provided with answers.

Level 3 Scope and sequence

Unit	Unit title	Strand	Learning objectives	Key words
1	Think maths	Think maths	• Establish a classroom environment conducive to thinking and working mathematically • Set up positive norms • Establish key messages of growth mindset mathematics	estimate, growth mindset, mindset mistake, strategy
2	Number and place value	Number	• Count forwards and backwards in ones, tens and hundreds from any number between 0 and 1000 • Identify, represent and estimate numbers using different representations • Understand that 10 tens = 100 • Recognise the place value of each digit in a 3-digit number (hundreds, tens, ones) • Understand and explain that the value of each digit is determined by its position in that number (up to 3-digit numbers) • Compose, decompose and regroup 3-digit numbers • Partition 3-digit numbers to show place value • Order and compare 3-digit positive numbers using > and < • Estimate numbers of objects/people up to 1000 • Recite, read and write numbers up to 1000 in numerals and in words • Round 3-digit numbers to the nearest 10 or 100 • Solve number problems and practical problems using number facts and place value	number names to 1000, place value, digit, hundreds, tens, ones, one thousand, place-value table, column, number sentence, difference, estimate, range, sum, partition, regular partition, irregular partition, number line, order, round, round up, round down
3	Length	Measure	• Measure, compare, add and subtract lengths in m, cm and mm • Use instruments that measure length in m, cm, mm • Estimate and measure length in cm, m, km • Understand the relationship between units • Round up and down to estimate and measure	length, width, height, centimetre, metre, millimetre, estimate, measure, round
4	Patterns and sequences	Number	• Use a shape or object to represent an unknown quantity in addition and subtraction • Recognise and extend linear sequences and describe the term-to-term rule • Extend spatial patterns formed from adding and subtracting a constant • Solve problems, including missing number problems, using number facts, place value, and more complex addition and subtraction	count on, count back, pattern, sequence, term, term-to-term rule
5	Lines and angles	Geometry	• Recognise an angle as a description of a turn • Identify right angles • Recognise that a straight line is equivalent to two right angles or a half turn • Identify horizontal and vertical lines and pairs of perpendicular and parallel lines • Recognise that two right angles make a half turn, three make three-quarters of a turn and four make a complete turn • Identify whether angles are greater than or less than a right angle	angle, right angle, half turn, full turn, quarter turn, three-quarter turn, square corner, 90-degree angle, straight line, parallel lines, perpendicular lines, horizontal lines, vertical lines
6	Polygons	Geometry	• Identify, describe, classify, name and sketch 2D shapes by their properties • Recognise angles as a property of shape • Differentiate between regular and irregular polygons • Draw polygons by joining marked points • Identify parallel and perpendicular sides of polygons • Identify right angles in 2D shapes presented in different orientations • Identify horizontal and vertical lines of symmetry on 2D shapes and patterns • Sketch the reflection of a 2D shape in a horizontal or vertical mirror line	polygon, side, vertex (vertices), square, rectangle, triangle, pentagon, hexagon, quadrilateral, octagon, regular polygon, irregular polygon, circle, half, quarter, degrees, protractor, compasses, set square, ruler, symmetry, line of symmetry, symmetrical

Unit	Unit title	Strand	Learning objectives	Key words
7	Addition and subtraction	Number	• Recall addition and subtraction facts that bridge 10 • Count on and back in steps of constant size • Understand and use commutative and associative properties of addition • Understand the inverse relationship between addition and subtraction, and how both relate to the part–part–whole structure • Estimate, add and subtract whole numbers with up to 3 digits (regrouping of ones or tens) • Estimate the answer to a calculation and use inverse operations to check answers • Add and subtract numbers mentally, including: a 3-digit number and ones; a 3-digit number and tens; a 3-digit number and hundreds • Add and subtract numbers with up to 3 digits using columns • Regroup 3-digit numbers, using hundreds, tens and ones • Apply place-value knowledge to known additive number facts (scaling facts by 10) • Recognise and calculate complements to 100 and complements of multiples of 10 or 100 (up to 1000) • Solve problems, including missing number problems and more complex addition and subtraction	addition, subtraction, strategy, add, subtract, count on, count back, total, sum, difference, fact family, hundreds, tens, ones, regroup, column, estimate, inverse operation, carry
8	Money	Measure	• Interpret money notation for currencies that use a decimal point • Add and subtract amounts of money to give change, using both dollars ($) and cents (c) in practical contexts.	money, cost, dollar, cent, local currency names, decimal point, placeholder, price, change
9	Mass	Measure	• Estimate, measure, compare, add and subtract mass in grams (g) and kilograms (kg) • Use instruments that measure mass • Understand the relationship between units	mass, weight, kilogram, gram, balance, scale, weigh, heavy/heavier/ heaviest
10	Multiplication and division	Number	• Understand and explain the relationship between multiplication and division • Understand and explain commutative and distributive properties of multiplication and use these to simplify calculations • Know 1, 2, 3, 4, 5, 6, 8, 9 and 10 times tables and related division facts • Estimate and multiply/divide by 2, 3, 4 and 5 for whole-number products up to 100 • Use knowledge of factors and multiples to understand tests of divisibility by 2, 5, 10, 25, 50 and 100 • Multiply numbers by 10 using knowledge of place value • Multiply 2-digit numbers by 1-digit numbers, using mental and formal written methods • Solve problems, including missing number problems, involving multiplication and division, including positive integer scaling problems and correspondence problems in which n objects are connected to m objects • Count from 0 in multiples of 4, 8, 50 and 100 • Apply known multiplication and division facts to solve contextual problems with different structures, including quotative and partitive division	array, times table, repeated addition, times, multiplication, multiply, multiple, division, divide, group, share, share equally, remainder, inverse operation, product, factor, divisible, divisibility
11	Perimeter and area	Measure	• Distinguish between perimeter and area • Understand that perimeter is the total distance around a 2D shape and can be calculated by adding lengths • Understand that area is how much space a 2D shape occupies within its boundary • Estimate, measure and calculate the perimeter of a shape using appropriate metric units and area on a square grid	perimeter, length, distance, sides, add, centimetre, area, square units, square centimetre, square metre
12	Data	Statistics	• Plan and conduct an investigation • Ask and answer non-statistical and statistical questions • Record, organise and represent categorical and discrete data • Choose and explain which representation to use in a given situation: Venn and Carroll diagrams; tally and frequency tables; pictograms and bar charts • Interpret and present data using bar charts, pictograms and tables • Interpret data, identifying similarities and variations, within and between data sets, to answer non-statistical and statistical questions • Discuss predictions and conclusions. • Solve one-step and two-step questions using information presented in scaled bar charts, pictograms and tables	table, row, column, tally mark, tally table, survey, bar chart, total, Venn diagram, element, overlap, intersection, Carroll diagram, category

Unit	Unit title	Strand	Learning objectives	Key words
13	3D shapes	Geometry	• Identify, describe, sort, name and sketch 3D shapes by their properties • Recognise pictures, drawings and diagrams of 3D shapes • Draw 2D shapes and make 3D shapes using modelling materials • Recognise 3D shapes in different orientations and describe them	3D shape, cube, cuboid, cone, cylinder, prism, pyramid, face, vertex (vertices), edge, triangular prism, square-based pyramid, net
14	Position, direction and movement	Geometry	• Interpret and create descriptions of position, direction and movement including reference to cardinal points	clockwise, anti-clockwise, quarter turn, half turn, three-quarter turn, full turn, position, above, below, under, next to, between, direction, up, down, left, right, forwards, backwards, cardinal points, north, south, west, east, grid, block, row, column
15	Fractions	Number	• Interpret and write proper fractions • Recognise, find and write fractions of a discrete set of objects: unit fractions and non-unit fractions with small denominators • Recognise and use fractions as numbers and fractions of a set of objects: unit fractions and non-unit fractions with small denominators • Count up and down in tenths • Recognise that tenths arise from dividing an object into 10 equal parts and in dividing 1-digit numbers or quantities by 10 • Understand and explain that fractions are several equal parts of an object or shape and all the parts together equal one whole • Understand and explain that fractions can describe equal parts of a quantity or set of objects • Recognise and show, using diagrams, equivalent fractions with small denominators • Compare and order unit fractions, and fractions with the same denominator • Add and subtract fractions with the same denominator (within one whole) • Solve problems involving fractions	fraction, half, third, quarter/fourth, fifth, sixth, tenth, numerator, denominator, equivalent fractions, greater than, less than

16	Capacity and temperature	Measure	• Estimate, measure, compare, add and subtract volume/capacity in litres and millilitres
			• Use instruments that measure capacity (ℓ/ml) and temperature:
			  ○ estimate first and then measure
			  ○ round up if the measurement is past or equal to the halfway point between readings
			  ○ round down if the measurement is less than the halfway point between readings
			  ○ use instruments that measure temperature
				capacity, full, half full, litre, millilitre, temperature, thermometer, degree, hot, warm, cool, degrees Celsius (°C)

17	Probability	Statistics	• Use informal language and the language of probability to describe the chance of an event happening • Conduct probability experiments, and present and describe the results	chance, possible, impossible, event, definite, likely, unlikely, equal chance, outcome, probability, experiment
18	Time	Measure	• Choose the appropriate unit of time for familiar activities • Read and record time using digital notation (12-hour clock) • Interpret and use information in timetables (12-hour clock) • Understand the difference between a time and a time interval • Find intervals between the same units of time in days, weeks, months and years • Tell and write time from analogue clocks, including Roman numerals from I to XII, and 12-hour and 24-hour digital clocks • Estimate and read time with increasing accuracy to the nearest minute; record and compare time in terms of seconds, minutes and hours; use vocabulary such as o'clock, a.m./p.m., morning, afternoon, noon and midnight • Know the number of seconds in a minute and the number of days in each month, year and leap year • Compare durations of events (for example, calculate the time taken by particular events or tasks)	time, o'clock, half past, quarter past/to, analogue clock, digital clock, Roman numerals, second, minute, hour, day, week, month, year, time interval, before, after, earlier, later

Section 1 Introduction

This Teacher's Book is designed to support the component parts of *Nelson Maths* Level 3.

Sections 1–4 of this Teacher's Book (pages 14–31) provide background and reference information, tips on classroom organisation, and a bank of activities and games to use for teaching practically. Section 5 provides lesson notes for the Pupil Book and Workbook unit-by-unit and page-by-page.

Strands and skills

In Levels 1 to 6, the content is divided into strands:
- Number – numbers and the number system, and calculating
- Measure – money, length, mass and capacity, and time
- Geometry – geometry, position and movement
- Statistics – organising, categorising and representing data
- Algebra – use of formulae, variables and equations.

In this course, problem solving is not treated as a separate strand but is integrated into the materials at all levels.

Fundamental principles

This series makes the following assumptions about the teaching of mathematics:
- Children need concrete (practical, hands-on) experiences in order to acquire sound mathematical understanding. Like adults, children learn best when they investigate and make discoveries for themselves.
- Children refine their understanding and develop conceptual structures by talking about their own thinking and what they have done.
- Individual children develop at different rates. While some will find certain elements of mathematics difficult, others will understand them quickly.
- Children learn in a variety of ways, so mathematics teaching should provide a rich and wide variety of experiences. Children need plenty of opportunities to apply what they have learnt and to relate their mathematics work to other areas of the curriculum and their daily lives.
- Children will become more mathematically able if they are allowed to develop reliable personal ways of working. The formal recording used by mathematicians comes with time and experience, and we cannot expect young children to understand it or use it themselves before they are ready to do so.

The conventions (formal methods) should be taught only once children are confident in their own knowledge, concepts and skills.
- Children learn mathematics most effectively when they enjoy what they are doing and when they can see the relevance of what they are learning.

This course reflects current thinking about the most effective ways of teaching and learning mathematics at primary level. It recognises the professionalism of the teacher, and acknowledges that teachers are the best judges of which experience(s) are most appropriate at different stages for the children in their classes. For this reason, this course does not impose a rigid, inflexible structure. Instead, it provides a wide variety of practical activities and games linked to clearly defined purposes and objectives. The teacher selects according to the needs of their classes, groups and individuals.

Individual differences and inclusivity

Everyone learns at their own pace, and in different ways. We also need to recognise that each child enters the classroom with their own experiences, their own identity, and their own hopes, fears and curiosities.

This course recognises individual differences and aims to give children the chance to explore the world of mathematics and solve problems in their own way. To achieve this, we believe that each child needs to feel that maths is for them too. Therefore, examples and illustrations need to include a wide range of children of different genders and ethnic, cultural and linguistic backgrounds and should not promote stereotypes.
- When giving children additional maths problems to solve:
  - Use examples of girls doing things that are traditionally considered 'for boys'.
  - Use examples of dads as the primary child carers as well as of working mothers and female leaders.
- Be aware of children's differences and acknowledge different bodies: not all children have ten fingers and ten toes, and some may not be able to walk or run as easily as others.
- Avoid comparisons of height or weight that some children may experience as body shaming.
- Celebrate slower workers in your classroom as much as faster workers. Praise reasoning and focused working rather than speed. Never focus on speed testing or 'who can find the answer the fastest'. Inclusivity means acknowledging that our slower, deeper thinkers are some of our best mathematical minds too. Maths is not about speed. It is about reasoning.

- Praise different ways of working. Some children need to move and fidget and play in order to solve problems. Include those who think visually or rely on practical manipulatives and drawings to solve their problems.
- Struggle and confusion are key parts of the process of maths learning. Never dismiss a child's struggle as evidence that they cannot do maths or that they are not mathematically minded.

- When you are planning classroom activities, bear in mind the abilities and needs of your group and change or adapt activities as necessary. For example, it may be more appropriate to count sets of ten counters, or use ten frames (see page 21) than to use fingers for counting. When activities expect children to identify, sort or match colours, adapt these activities for children who have colour-blindness, by focusing on pattern rather than colour.

Section 2 Teaching approach

A growth mindset

In recent years, scientists have made important discoveries about how the brain learns and grows. A key discovery is the notion of neuroplasticity or brain plasticity. Brain plasticity is the brain's ability to change connections and neural pathways. Put simply, this means that the brain can learn and grow throughout a person's life.

These discoveries have led many educators to talk about 'growth' versus 'fixed mindset'. People with a fixed mindset believe their abilities are determined and fixed, for example, by genetics or by factors outside their control. A fixed mindset can lead to fears of making mistakes. Children with fixed mindsets may believe that they are either 'good at' or 'bad at' maths, and may easily get frustrated when they find problems challenging. We need to help children to understand that our maths ability changes all the time as we learn and grow. By getting children to understand and believe that growth, development and improvement are always possible, we can foster a 'growth mindset', in which children are not afraid of tackling difficult problems.

Getting rid of maths myths
Every child has the potential to learn maths, irrespective of their existing level of mathematical ability. The ability to learn maths is also not related to other traits such as gender, race or ethnicity. There is no such thing as a 'maths person'. It is extremely important that teachers understand that all children are able to learn to think mathematically. Maths is not a special subject and does not require any special natural 'talent'. Some children may grasp concepts more quickly than others – just like in other subjects – but giving them the time to develop their understanding allows all children to achieve highly in mathematics.

The importance of mistakes
Scientists have also studied the brains of people attempting to solve problems. They found something that might seem surprising – that more neural connections and pathways form when we make mistakes than when we get things right! Although getting correct answers is usually praised by teachers, it is key that we also start understanding the importance of mistakes. Mistakes offer a valuable opportunity to explore why and how a concept makes sense. You can explore mistakes in a productive way with your class by:
- frequently reminding the children that making mistakes shows we are learning something new
- ensuring that the children know that mistakes (and questions) are welcome in your classroom
- discussing mistakes with interest and curiosity
- instead of always asking a question that invites a correct answer, sometimes phrasing questions to ask the children to identify the *incorrect* answers and then explain why they don't work.

You can read more about growth mindset in books by Carol Dweck and Jo Boaler, or search for these online.

The do–talk–record model

The learning framework for this course can be summarised as: *do–talk–record*. In other words, the starting point is always to explore and investigate concepts. Talking, discussing and explaining follow. Finally, written work can record and represent the concepts that the child has already explored practically and in discussion.

Doing
The first step in developing and understanding concepts is to let children:
- handle and manipulate apparatus
- play games
- investigate patterns and rules using real objects
- model situations and problems using real objects.

Children need time to play and explore in this way before they are expected to communicate about their work.

Talking
Children can make sense of what they have been doing by discussing:
- what they have done
- why they have done it
- what they have found out.

This allows them to generalise concepts and ideas from their particular experiences. The teacher's role is to create situations for discussion and to ask open-ended but directed questions. Most of the activities in this Teacher's Book will help you to facilitate discussion and will encourage the children to listen to each other and experiment with different ways of thinking about and solving problems.

Recording

At Level 3, children are not likely to have refined the skills and knowledge or developed the use of strategies for solving problems. They will need to use informal and very personal methods (jottings) of recording steps in a process, or keeping track of what they have done.

Jottings are an important step in moving towards non-standard methods of calculation (such as diagrams and jumps on a number line). These will give the children a foundation for more concise standard written methods of recording. It is very important that you allow, and in fact encourage, the children to make use of jottings as they work. Here are some possible ways of doing this in the classroom:

- Do jottings of your own as you work out solutions. For example, if you are demonstrating how to add $7 + 8 + 3$ you might jot the following on the board to show your thinking:
  $7 + 3 \rightarrow 10$
  $10 + 8 \rightarrow 18$
- Talk through the jottings as you make them. For example, say: *It's easier to add tens, so I'll add 7 and 3 first.* This modelling process helps the children to see that jottings are important and useful.
- Make space for jottings in the children's exercise books. You can reinforce the importance of jottings as a means of showing your working by encouraging the children to jot as they work. If you only allow jotting on scrap paper, the children may think it is not as important or valuable as the 'real' work in their book.
- Do activities where jotting is the point of the activity. For example, ask the children to represent a calculation visually in as many ways as possible, or ask them to work out problems where they will need to jot down steps to keep track of the process (such as *How many different ways can you pay for an item costing … using coins only?*).
- Ask the children to share their jottings and compare them to show that there are different methods of working. This can help the children to see that some strategies are more efficient than others, which can refine their own thinking. For example, in the coin task above you may find that some children draw coin combinations, others list them and those who are more able and confident may make a table and work more systematically. All of these methods may provide the correct answers, but obviously some will take longer than others.

- In the early stages of using apparatus in a new way, recording may take the form of drawings or words and drawings. Some children will gradually find this time-consuming and will simplify their recording independently. Others may need your suggestions and encouragement. As a teacher, you will need to work out carefully when a child is ready to use a standard mathematical symbol or format, so that recording is based on full comprehension. It is crucial that you do not force the children into formal and standard methods of recording calculations before they have fully grasped the process and are confident in the methods.

Exploring and investigating

Traditionally, primary mathematics has almost always involved short, directed tasks that result in 'right' or 'wrong' answers. This course provides a balance between short, fairly self-contained activities and open-ended investigations that sometimes require the children to collect information, make their own observations, ask questions or complete activities at home. Most of the activities are designed to develop children's awareness of the range of mathematical possibilities open to them when they are tackling a mathematical task.

It is important for the children to experience a range of different kinds of questions and tasks, and to develop the understanding that mathematics gives us tools to investigate and test our answers to real-life questions. In most cases, there is more than one way to solve a question. Even a simple question, such as $10 - 8$, can be solved in a range of ways: by working out the difference between 10 and 8, counting on from 8 to 10, using a number line or a chart, breaking the 10 into 2 fives and the 8 into a five and a three, and so on.

As much as possible, allow the children to take control, make decisions and explore the many avenues that can arise from a simple starting point. Always encourage children to ask 'What if … ?' and 'Why?' when investigating. These questions may lead to new challenges, fresh understanding and the development of new skills.

Many investigations have no final 'answer' or easily accessible generalisation for the children. Some have a simple pattern or rule that may be discovered and explained. However, many children will want to know why certain patterns repeat, and offer explanations about the rules that govern them. This is the first step towards generalisation; encourage this by asking questions, for example: *Why is the same number added each time?* or *Can you guess what will happen next?*

The value in investigations is in children following them as far as they can, and in the new skills that they acquire on the way. For some children, the early, often concrete, experimentation is enough to give them confidence, and increase their enjoyment of using already acquired skills.

Use investigations to introduce new concepts. For example, introduce number patterns by presenting number chains (sequences), and letting the children work out what comes next or what is missing. Introduce geometric patterns through explorations of colour arrangements using pattern blocks, tiles or geoboards. The children can explore the relationship between 2D shapes and 3D shapes by identifying the 2D shapes that are the faces of the 3D shapes.

As the children develop an investigative approach, help them to work systematically, and remind them of the question they are working to answer. Encourage making new observations and discoveries along the way, but ensure that these do not eventually distract from the question the children are working on. Ask questions such as: *What does this show us? What do you notice? What do you wonder? How can you use this to work out the answer? What other information do you need? How can you find it?*

Many everyday objects can provide rich sources of investigative work. The 100 chart, addition square and multiplication square all contain many fascinating patterns. Children can also explore patterns in shapes and solids, such as the relationships between sides, corners and angles in 2D shapes and between faces, edges and vertices in 3D shapes.

Number talks

Number talks are open-ended, meaning they present a question that has multiple possible answers, and multiple possible strategies for solving.

Children can use agreed hand signals as shown below, instead of calling out or putting their hands up. However, these may not be culturally appropriate in all countries, and you may want to come up with alternative signals that are suitable for your class.

Step 1: Present the problem and give time for independent thinking
First present the problem, then ask the children to think to themselves quietly how they could solve it. They can use hand signals to show when they have an idea.

Note that when children signal that they have an idea, you can encourage them to think of more ways to solve it. They can indicate the number of ways by raising more fingers, as shown.

Step 2: Share in pairs
In pairs, children share their suggestions for how to work a problem out. Note that explaining how they solved it is as important as (or more important than) the final answer. They can use a hand signal to indicate they are finished, such as a fist to the chest.

Step 3: Children share their strategies
Ask individual children to share their strategies with the class. Discuss each strategy as the child demonstrates for the class. Continue using hand signals so that the children can engage with the discussion and take turns to share their own working. The children can hold their thumb and little finger out for 'I agree' or 'Me too'; hold their hands with palms down and move them in a criss-crossing motion for 'I respectfully disagree' and raise a finger for 'I have another way to do it'.

Step 4: Discussion
The discussion will take place as other children show that they agree or disagree, and offer alternative strategies.

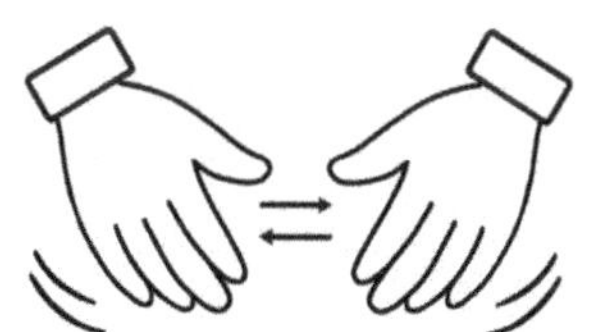

Principles of the number talk strategy
The number talk strategy aims to get all children to engage actively and creatively with mathematics. To achieve this engagement, it is important to remember the 'big ideas' behind the number talk:

- *Value the contributions of all children.* Throughout the discussion help them to articulate their thoughts and reach a clearer understanding of how they are working things out. The emphasis is on the process and understanding, not on reaching the right answer fast.
- Children often answer in a roundabout way, with a very general idea or 'sense' of what they want to say. Use questions to *clarify, refine and explore* the meaning behind each response.
- Encourage children to explore *conceptual explanations*, rather than procedures.
- Gently *explore interesting mistakes* because these provide opportunities for deeper understanding. We can often better understand how something does work by reaching a clearer understanding of how it doesn't work.

- Praise children for effort, for making sense of a problem and for *persistence in the face of difficulty*, rather than for efficiency. We like to tell children that if it is difficult, that tells us we are in the place of learning. If it is easy, that means we've already learnt it and it's time to find more challenge.
- Create a *safe community space* where children feel they can share their ideas without criticism or judgement.
- Encourage *sharing ideas*, rather than competition. Children should understand that learning comes from connecting with other people's ideas and building on them.

Finding out more about number talks

It is important to remember that number talk is a format or technique, not a specific lesson or set content.

- View number talks in action, shared by teachers, on the Internet.
- Find posters and strategies for number talks through an online image search.
- Books on number talks include: *Number Talks: Whole Number Computation, Grades K-5: A Multimedia Professional Learning Resource* by Sherry Parrish; *Classroom-Ready Number Talks for Kindergarten, First and Second Grade Teachers* by Nancy Hughes.

Numberless word problems

Many teachers fear problem solving because they present a problem, and then watch as the children try to guess which operation to use, or which numbers to arrange into a number sentence. Numberless word problems offer a useful strategy for encouraging children to think and understand questions before looking for the 'maths answers'.

The numberless word problem is presented with no numbers at first, and the numbers are gradually added. The numberless word problem does two things:

- It removes the numbers.
- It removes the question.

This forces the children to slow down and think about the problem before they can attempt to solve it.

To present a numberless word problem, follow these steps:

- **Step 1:** Present the word problem you want to use, but remove the numbers and the question. Use words such as *some, more, joined, went away, took away, fewer,* and so on. Ask the children to tell you what is happening in the story. Here is an example to practise addition:

> There are some children in the playground. Some more children come out to play. Now there are many children in the playground.

- **Step 2:** Ask the children to think about what the question could be in this story. Allow plenty of thinking time and invite the children to be 'question detectives', guessing from the clues what the question is going to be. They may have a variety of suggestions, such as: *How many were in the playground to start with? How many children came out? How many were there altogether?* Once they have identified the possible questions, you can reveal the question, but still without numbers:

> There are some children in the playground. Some more children come out to play. Now there are many children in the playground.
> How many children came out to play?

- **Step 3:** Ask the children to identify what information they need in order to solve the problem. First, ask the children if they can answer the question. They cannot – because they don't have the information they need. Let them suggest what they need to find out in order to solve it. Give them the first piece of information they need, by crossing out 'some' and writing a number in:

> There are ~~some~~ 15 children in the playground. Some more children come out to play. Now there are many children in the playground.
> How many children came out to play?

Again, ask if they can answer the question. When they say no, let them work out what other information they need to solve it. Let them work out what else they need to know. Finally, cross out 'many' and change it to a number (24).

- **Step 4:** As with number talks, give the children time to solve the problem using their own strategies. Follow the steps of a number talk as you let them share and discuss their strategies.

Mathematics in real life

Some children may struggle to understand the relevance of mathematics in their everyday lives. This course places great emphasis on making children aware of the relevance of mathematics in real life.

In this Teacher's Book, you will find ideas for using the child's own environment as a stimulus for mathematical activities. The Pupil Book and Workbook frequently require the children to look at the mathematics in the classroom, the playground and their own homes. Each set of activities and problems requires new skills and fresh understanding. Many questions are open-ended or have no exact solution, and the children are asked to make predictions, generalisations and estimates, and to evaluate their own answers. Encourage these skills in all areas of the curriculum. Children use their understanding of mathematics at home and at school, in situations such as sorting toys or other items, telling the time, looking for and making patterns, helping to prepare food, and playing board and other games.

In school
In school, there are many opportunities for you to teach mathematics through familiar situations, so that the children understand why it is useful and appreciate the order and sense that mathematics gives to life. For example, the children can identify the date each day, as well as the time at various points throughout the lesson. Registration, dinner money, timetables, sorting and putting away equipment will provide a range of relevant experience in data work, measures, shape and space as well as number.

In play
Children of all ages should have opportunities to play both in and out of school. This offers them the freedom to explore new situations, to make discoveries for themselves and to be creative. Unfamiliar mathematics equipment should be introduced through play, with the children exploring the functions and possibilities of the materials. A good example of this is a geoboard – you can show the children how to use it, and then let them work out what different kinds of shapes they can make using the elastics and the board.

Construction kits offer children the opportunity to explore shapes and inverse operations, through building and dismantling. Note that for any materials that are unavailable at your school or in your area, you may be able to find an online version.

At home
Part of the teacher's role is to involve parents and guardians in the children's learning and there are many opportunities for children to develop their mathematical understanding in simple ways at home. However, in order for this to happen, you may need to give parents some guidance about how easy this can be.

Many parents lack confidence in their own mathematics ability, and have the idea that maths is something complicated that children need to learn at school. A useful technique is to offer a 'learning café' – a short presentation for parents where you present simple ideas that they can use at home with their children. If time is an issue and parents cannot attend a session like this, adapt it to a page or booklet of easy home ideas for supporting mathematics. You can do this by pointing out the contexts in which we can use mathematics at home. Here are some examples:

- *Counting:* The children can practise counting through simple games and when using building blocks. Shopping is also a great opportunity for counting. Parents can ask their children to count out given numbers of items, or identify how many things are in a bag or box (how many eggs in differently sized trays, how many rolls in a bag, how many oranges in a container, and so on).
- *Patterns:* Show parents how to make patterns out of construction blocks. They can also talk to their children about everyday patterns that they notice in the world around them.
- *Shapes:* Explain to parents that identifying and sorting shapes and everyday objects is a key skill. Naming and identifying shapes around the house can help consolidate this skill.
- *Ordering:* When tidying and ordering the home, we use the skills of sorting things, arranging from largest to smallest, identifying what belongs where. These are all skills that support mathematical reasoning. Encourage parents to involve their children in these activities.
- *Measures:* When we cook, we measure ingredients, either with scales or with cups and spoons. Parents can show their children how to use these.
- *Time:* Draw attention to telling the time, comparing how long things take, understanding minutes and hours as well as times of day, and so on.

You can find additional parent notes for each level on Oxford Owl. See page 9 for more details.

Section 3 An environment for exploration and play

Organising the space

All teachers have their own preferences about how best to organise the available space. However, here are some useful guidelines for any classroom.

Storage

Always store equipment so that the children have easy access to it and you can check it periodically. Label all items clearly and encourage the children to make their own decisions about what they need. From the very beginning, insist that the children pack up and return equipment to the correct storage containers or shelves.

A mathematics centre

Some teachers may choose to have a dedicated area, maybe a corner, of the classroom where they keep maths materials such as counting frames, ten frames and Numicon number frames, prepared number game boards, geometric shapes, and so on. You can also display posters about shapes and numbers, patterns and other topics as these come out of the weekly lessons. Other teachers may prefer to integrate this into the rest of the classroom.

Recycled resources

Many household items can be repurposed to make mathematical equipment. Invite the children and their families to collect some of these materials at home in order to keep your classroom well resourced:

- bottle caps
- egg boxes
- clean glass jars
- buttons
- plastic caps from juice or milk cartons
- plastic yoghurt pots
- beads
- natural objects such as shells, seed pods and stones for the children to arrange in size order.

Store the materials neatly in plastic containers or jars.

Individual work, classwork and groups

Whole class teaching

Often you will work with your whole class, especially at the beginning and end of a lesson. The course offers plenty of ideas for this kind of approach. The children may be seated together on chairs or on the floor, arranged in a circle for a discussion or clustered around you for a demonstration. If you are calling on individuals to show things to the group, make sure there is enough space at the front for them to come up, and make sure everyone can see and hear the volunteer child's answer.

Set up some classroom rules at the beginning of the year and go over these at the beginning of each term or even each week if necessary. Classroom rules might include listening when others are speaking and waiting turns. Many teachers find it useful to use signs for *agree*, *disagree* and *I know the answer* rather than having the children raise their hands or shout out. The traditional approach of raising hands can be intimidating for quieter children or those who take a little longer to think about their answers. Hand signals allow teachers to give more time to those in the group who need it. The illustrations on page 17 show some commonly used hand signals.

Group work

You can group the children in similar or mixed-ability groups, to suit the purpose of the work. This gives the children opportunities to collaborate and to discuss their work with each other and with you. It allows peer teaching to take place and for the work to be matched to their needs. It also allows you to work simultaneously with a number of children and this minimises the need for repeated explanations to individuals. Group teaching is an effective form of classroom organisation for both teacher and children.

Working individually or in pairs

At times it may be appropriate for the children to work as individuals or in pairs, to provide extra help to children who need it, or to stimulate and challenge children who have grasped a concept quickly. Working individually gives the children the opportunity to concentrate on their own thinking, to develop this through investigations and problem solving, and to experiment with materials. Children working in pairs have the opportunity to develop collaborative skills, to play games together and to share ideas in an investigation.

Using manipulatives

Manipulatives give children an opportunity to experience the use of mathematics through concrete materials and objects. There are suggestions for different kinds of manipulatives in the unit-by-unit teaching guidance section of this book.

Counting apparatus
Interlocking cubes

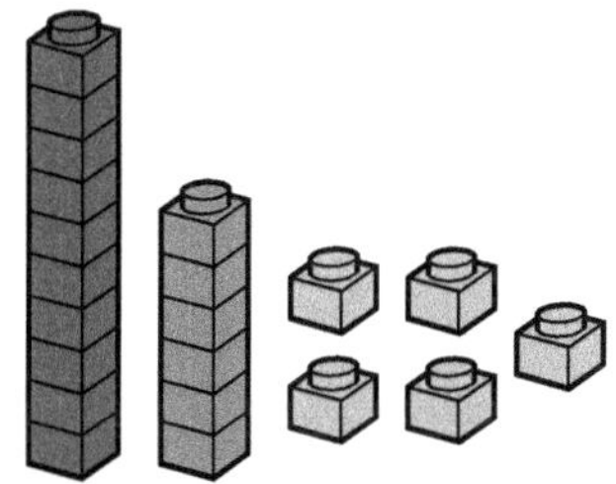

Interlocking or 'snap' cubes

Interlocking cubes are an invaluable resource for learning to count, especially as the children learn their number bonds up to 10, and then begin to work with tens and ones as they work with numbers up to 100.

Rekenreks

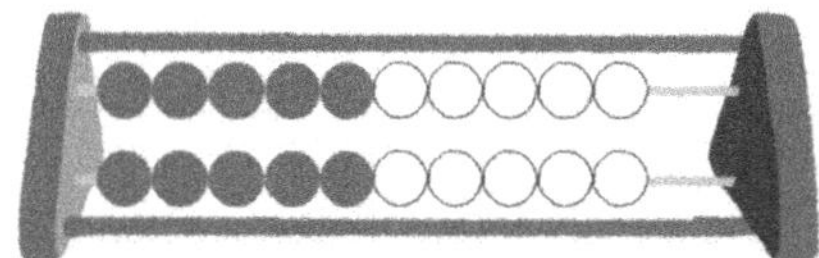

A rekenrek

The rekenrek was designed by a Dutch mathematics researcher to support children in developing their number sense in a natural, intuitive way. 'Rekenrek' means counting rack or counting frame. At the primary level, the device consists of two horizontal bars, each with ten beads. The beads are in two sets of five, shown by different colours (usually red and white). The rekenrek looks a little like an abacus, but the bars do not represent place values.

When we work with a rekenrek, we start with all the beads pushed over to the right, in the 'resting' position. To count or represent a number, we move beads over to the left. The rekenrek can be used for a range of skills, including:

- developing subitising skills within 5 and within 10
- counting to 10
- addition and subtraction
- composing and decomposing numbers up to 20
- recognising doubles and near doubles
- skip counting.

Working with a rekenrek allows children to develop a range of strategies intuitively.

If your school does not have access to commercially available rekenreks, you may be able to make some simple ones with dowel sticks and beads. You can also search for an online interactive rekenrek app.

Ten frames
A ten frame is a 2 × 5 rectangular grid, used most commonly with counters of one or more colours.

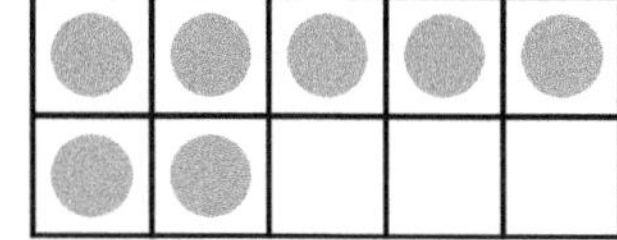

Using a ten frame to represent 7

Ten frames help children build and visualise numbers to 5, 10, 20, and 100. They can use the frames to count, represent, compare and calculate, working within a specific number range.

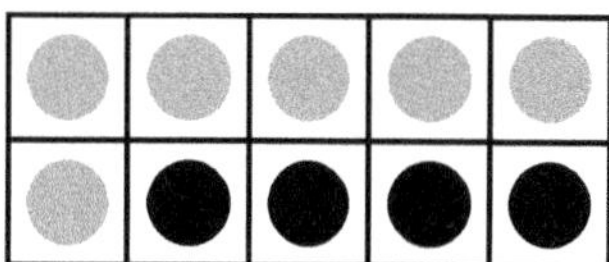

Using a ten frame to represent 6 + 4 = 10

Ten frames help children to see equivalences between quantities. So, for example, they will quickly see that 10 is equivalent to 2 fives, and that 20 is equivalent to 2 tens, or to 4 fives. This helps them move away from linear, one-by-one counting and develop a range of counting strategies.

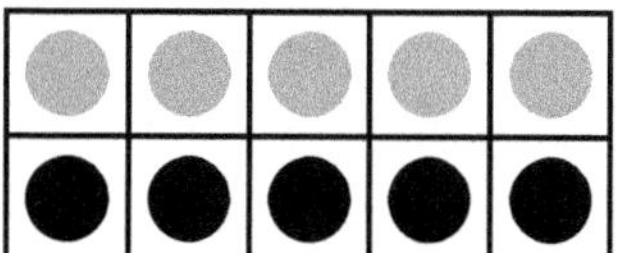 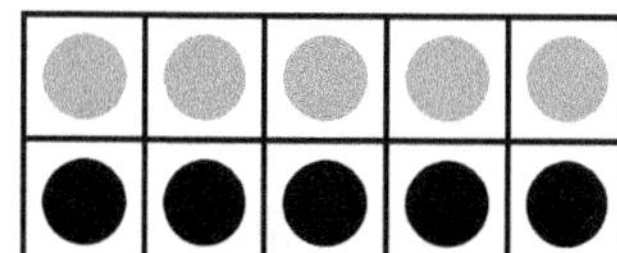

Using ten frames to represent 20 as 2 tens or 4 fives

Your school may have access to wooden or plastic physical frames. If you don't have these at your school, you can easily make your own ten frames and use sets of counters (five of one colour and five of another).

You can also search for interactive ten frames online.

Numicon (number frames)

Numicon

Numicon is another powerful and versatile concrete resource that supports children to visualise and work with numbers. The holes in the Numicon number frames represent numbers.

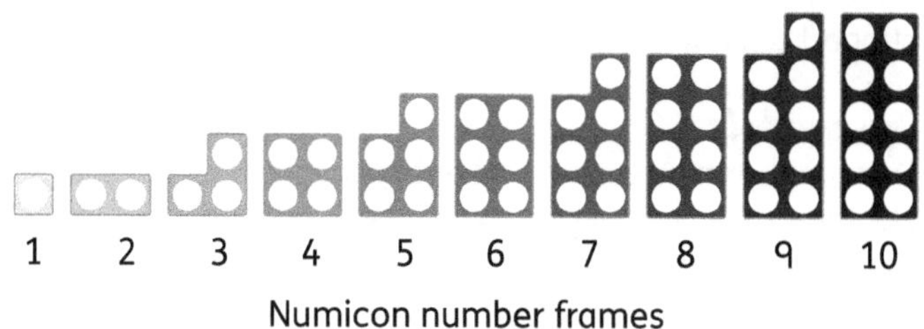

Numicon number frames

As with ten frames, the children can also use the Numicon number frames to count, represent, compare and calculate. When thinking about 5, for example, they can quickly see that 5 is 1 less than 6, but 1 more than 4. If they compare a 3- and a 6-number frame to a 9-number frame, they can see that 6 and 3 make 9.

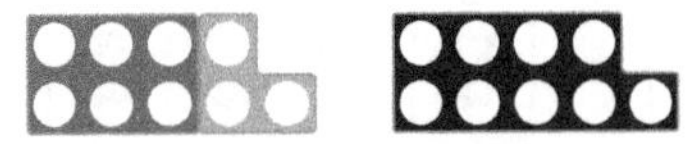

Using Numicon to represent 6 + 3 = 9

For more information on using Numicon number frames to teach mathematics, and the full range of Numicon products, visit https://global.oup.com/education/content/primary/series/numicon

Base-ten blocks

Base-ten blocks are a concrete resource made up of blocks of varying sizes. Each block can represent a different value. For whole numbers, they represent:

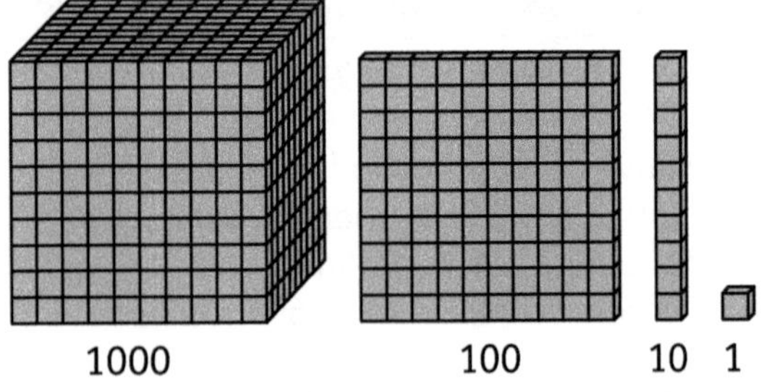

1000　100　10　1

They can also be used to introduce decimals, representing, ones, tenths, hundredths and thousandths respectively.

Like ten frames and Numicon number frames, base-ten blocks help children to represent numbers, compare and calculate. They are particularly effective when used to support partitioning, for example into tens and ones.

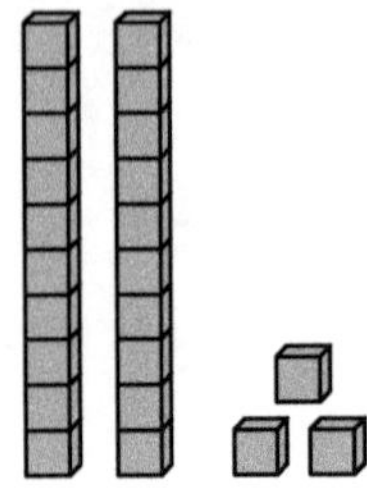

Representing 23 as two tens and three ones using two 10-rods and three ones cubes

Using base-ten blocks and place-value tables helps children to see how numbers change when multiplied and divided by powers of ten. Subtracting using base-ten blocks is a great, 'hands-on' way to introduce and support exchanging.

Bar models

Drawing bar models is excellent problem-solving strategy that children can use in both primary and secondary school. A problem that may otherwise seem very challenging can become more accessible when you 'draw' it.

Children should be encouraged to interpret bar models as well as draw their own when solving word problems.

There are two main types of bar models: the part–part–whole model and the comparison model. Which you choose to use will depend on the problem you are trying to solve.

For example, Tyra has 32 and she give 13 to Esme. How many marbles does she have left?

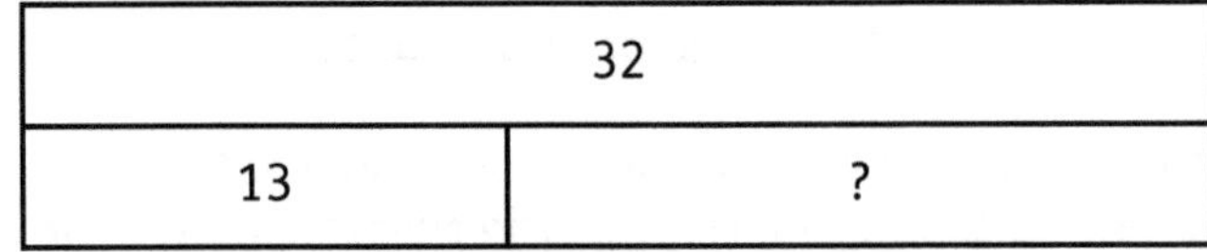

A part–part–whole bar model

Lewis completed 142 laps of the running track in one week. Joshua completed 52 laps in the same week. How many fewer laps did Joshua complete than Lewis?

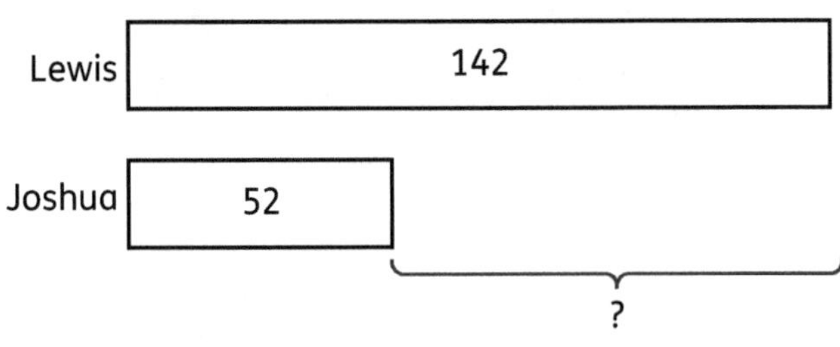

A comparison bar model

Place-value cards

Place-value cards have an 'arrow' or point on the right-hand side. Children can organise the cards horizontally or vertically to represent numbers in expanded notation. They can overlap cards and line up the arrows to form multi-digit numbers.

If any children have not previously worked with place-value cards, you will need to teach them how to use them. Begin by pointing out the arrows on the cards. Explain that these arrows always go on top of each other when you are making a number.

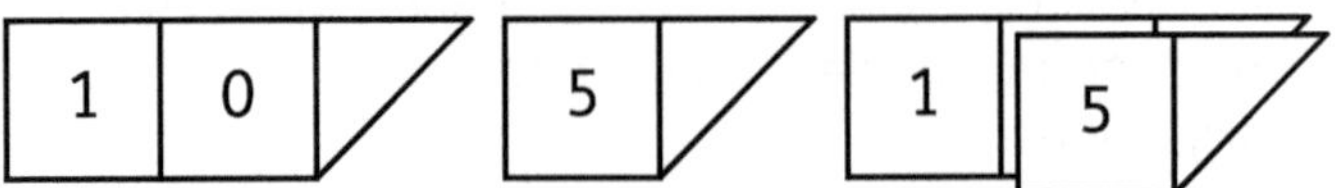

Place-value cards are an important teaching and learning resource and it would be useful to have a set available for each child. If possible, laminate the cards to make them more durable. (If you are making a set for each child, you may like to send the cards home for parents or carers to cut out.)

At Level 3, the children work with numbers to 1000, so you need to prepare hundreds, tens and ones cards. At higher levels these can be extended to as many places as needed and also to the right to show decimal places.

Here is a basic set of hundreds, tens and ones place-value cards:

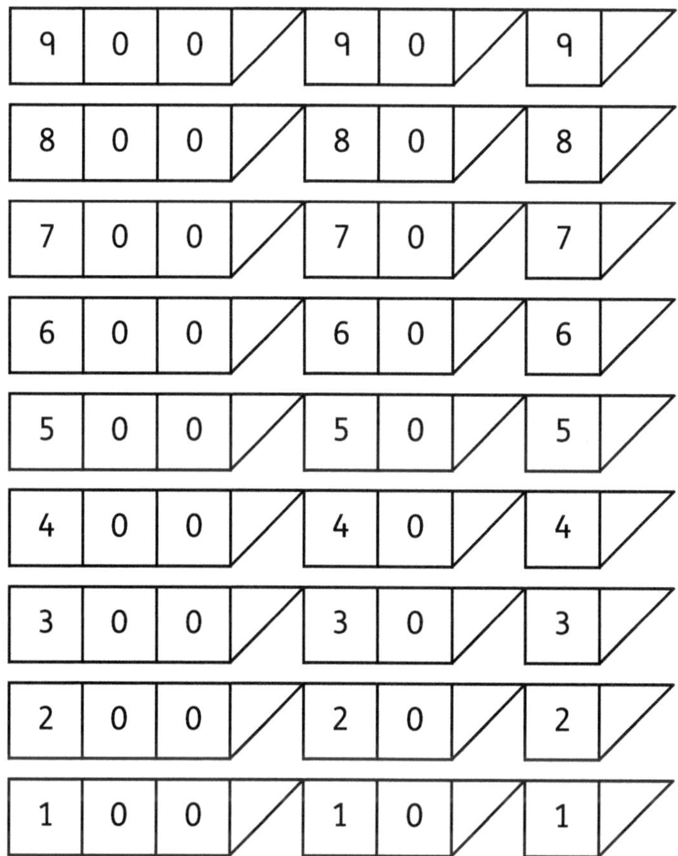

Place-value tables
Children can use place-value tables to build numbers using counters or blocks. If possible, make place-value tables on card and laminate them to make them more durable.

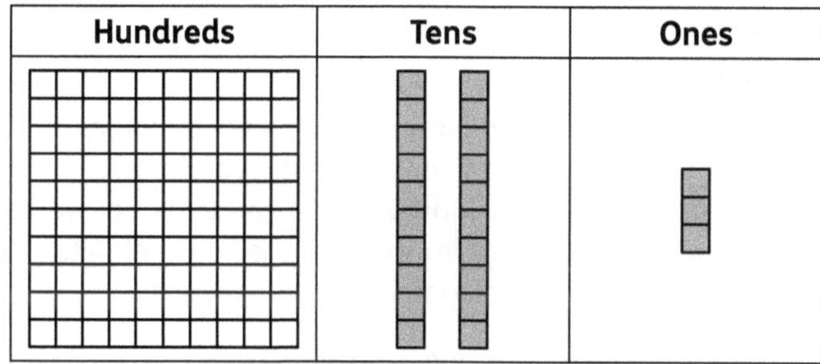

Hundreds	Tens	Ones

At Level 3, place-value tables need Hundreds, Tens and Ones headings. At higher levels these can be extended to as many places as needed and also to the right to show decimal places.

Materials for sorting and matching activities
Containers
Provide a range of containers of different sizes and shapes, and access to sand and/or water. Allow the children to fill and pour sand/water from one container to another. Ask questions and use the activities to develop the basic vocabulary of mass and capacity – *heavy, light, heavier, lighter, full, empty, nearly full, nearly empty, holds the same, holds a lot, holds a little, holds more than, holds less than,* and so on. These activities develop hand–eye coordination skills, as well as vocabulary, and also help the children to develop the idea of conservation of measure (for example, to learn that a wide flat container may hold the same as – or more than – a tall, thin container).

Feely bag
Prepare a 'feely bag' for the classroom. You will need a range of different items (for example, plastic cutlery, pens, rulers, buttons, combs, keys, coins, feathers) and a bag that you cannot see into. The children take turns to put their hand into the bag and select an item. They should hold the item and describe how it feels it to the class (*It is big, it is soft, it has smooth sides*). The others try to guess what it is. After some discussion, let the child remove the item and show it to the class. They should then describe it using its characteristics (including colour, size and shape). For example, the child might say: 'This is a comb. It is brown, and long and quite thin. It has teeth on one side and a flat handle for holding it.' You can include matching in this activity by drawing pictures of the items in the bag, or tracing around their outline. Ask the child to match the chosen item to the correct picture or outline.

Collections
Arrange the children in groups. Give each group of children half an egg box (or any other container with a number of compartments). Ask them to walk around the classroom or take them out into the school grounds and ask them to collect six items that are different from each other. When they have collected six items, the groups take turns to describe the items and say how they are different. For example, a group might say: 'This is a big brown seed. It is hard and rough on the outside. This is a small green leaf. It is smooth and round.' You can vary this by giving different instructions, for example: *Find six items that are all green, but all different in some way.*

Everyday objects
Matching pairs of everyday objects helps children to develop the idea of one-to-one correspondence. For example, ask the children to physically match items such as:

*   straws to bottles or cups
*   children to chairs
*   bags or coats to hooks or lockers
*   children to pencils or books
*   buttons to buttonholes
*   hats to heads
*   socks to shoes.

Provide crayons in different colours and ask the children to match them to counters using colour as the matching characteristic. Use this activity to teach the names of the colours if necessary. You can vary this to match any items using colour.

> With all activities using colour, if you have colour-blind children in your class, adapt the activity to use patterns, or combinations of colours that are within their visual range.

Shape jigsaws

Prepare a set of 'shape jigsaws' for the classroom. For each 'jigsaw' you will need two sheets of card the same size. Divide the sheet up into shape pieces like this:

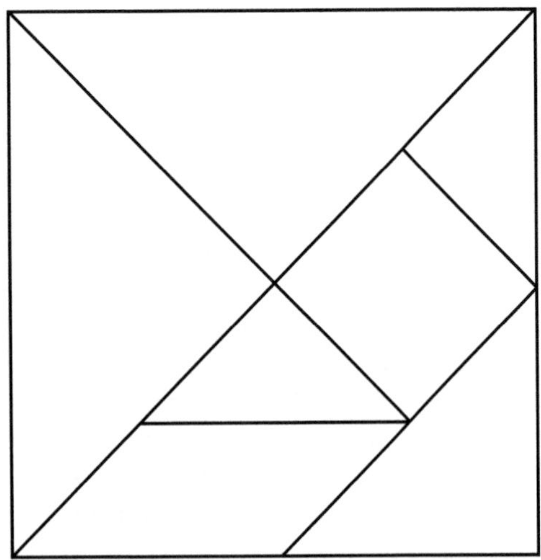

This shape jigsaw is based on the traditional tangram game. However, you can vary the shapes to make different puzzle boards.

Draw the same shape lines on both sheets. Leave one sheet intact. This is the guide board for the children. Cut along the lines of the other sheet to make a set of shapes. Store each puzzle (the guide board and the set of shapes) in a zip-lock plastic bag or large envelope. Give the puzzle to individuals or pairs of children and let them make the jigsaw by placing the cut-out shapes on their matching outlines. Note that the names and attributes of the shapes are not the focus of this activity, rather the point is to encourage the children to match like-to-like shapes.

Paint strips

You can prepare a fun matching activity using the paint colour strips you get from hardware or paint shops. You will need two of each strip for the activity. Use, for example, a strip showing shades of green. Leave one strip intact (or cut out the different shades and paste them in a random order on another sheet of card). Cut the other strip into pieces so that each piece shows one of the shades of green. Store each puzzle in its own zip-lock bag or envelope with the colour written on the outside.

Arrange the children in small groups and let them play with the puzzles. They should display the fixed colour strip (or sheet) and have the other pieces face down. They then take turns to pick up a piece and match it to the correct colour on the game board. To extend the activity, ask the children to identify items in the environment that match each shade of green (for example, 'The leaves on the tree outside are a dark green, like this strip.' 'The paint on the classroom wall is a very light green like this strip.').

Interlocking cubes

Prepare sets of interlocking cubes of different lengths (from one to about ten cubes). Cut lengths of wool or ribbon to match the length of the sets of cubes. Ask the children to match the lengths by comparing and then laying the wool next to the corresponding set of cubes. Extend this by including some pieces of wool that don't fit. Encourage the children to talk about why these don't fit. For example, they might say: 'This piece is too long for all the sets of cubes.' 'This piece is too long for this one, but it is also too short for this one.'

Egg boxes

Prepare a matching activity using egg boxes. Draw or stick one to four stars in each compartment. Ask the children to sort beads or other small objects to match the number of stars (one-to-one correspondence). They do not need to know the numbers or be able to count to do this.

Numbers, dots, objects and cards

To develop number sense, include matching activities that involve numbers of items arranged in different ways. For example, prepare cards with one to five items (shapes, dots, squiggles, fruits). Ask the children to make pairs of cards with the same number of items. Once you have taught numerals and number names, you can extend this activity to matching the amounts and then matching those to numerals and sorting them in order from fewest to most or vice versa.

Collect pictures of numerals (from 1 to 10 only) in different contexts and styles from magazines, birthday cards, calendars and product labels. Stick the individual numerals onto card and keep them in a box. Give the children a handful of cards and ask them to sort them into groups of symbols that are the same. They do not need to know the number names to do this, they can simply compare the shape and form of the numerals to match and sort them.

Games

A game is different from a teaching activity because there is an element of luck and chance in a game, sometimes leading to a winner. Some games are played alone, and children try to improve on their own results, while other games are played in pairs or groups.

Make sure the children understand the rules of the game before they start playing. If they don't, they may not play the game properly and this can result in conflict with others. The children should learn to:

- take turns and wait for their turn
- move their own game pieces only
- help each other if they can.

The children also need to learn that in games involving luck, players don't win or lose on the basis of ability. Explain that those players who get the best luck in a particular game win; those who lose were just less lucky. Winning graciously, as well as losing and playing fairly, are important life skills.

When the children play games in class, do not keep a record of who wins and who loses. If the children like a game, encourage them to play it in their free time and to take it home and play with family members if possible.

Section 4 Activity bank of warm-ups and mental maths

This activity bank includes a range of mental maths activities, as well as some suggestions for support and consolidation activities. Most of the ideas are for number work and operations, although there are also some suggestions for the other strands. You can use this as a resource for activity ideas for your class. The unit-by-unit lesson plans refer back to specific activities in this section as unit openers, suggestions for extra support and consolidation.

Try to include 10 minutes of mental maths activity each day. This section includes many examples that you can use as they are, or adapt to suit your own classroom. It provides a range of different types of activities (factual recall, games, grids, tables, problem solving and puzzles) to show some of the ways in which you can approach the mental maths part of the lesson. However, this is not a definitive list and some activities will appeal more to some classes and teachers than others. If you need additional ideas and suggestions, search for mental maths warm-ups online.

> Avoid any games that use countdown clocks or timers which turn mental maths into a race or speed test. Speed testing can be damaging to children's mathematical development and cause maths anxiety.

Number work and operations

Physical counting warm-ups
At Level 3, use these activities for skip counting in 2s, 3s, 5s, 10s, 20s, 50s and 100s.

- **Skipping rope counting:** Bring a skipping rope to class. Give individuals a chance to try to skip continuously to the highest number. The whole class chants the count as they go. Make sure as many children as possible get a turn.
- **Ball throw counting:** The children stand in a circle and throw the ball across the circle to a new person each time. How many throws can they keep it going for before the ball drops?
- **Focus count:** This is a quieter activity. The children stand in a circle. One child says 'one', another says 'two', another says 'three', and so on. The speakers should go in a random order (not around the circle in the order in which they are standing). So, children need to 'sense' when it's their turn. The challenge of the game is that two children must not speak over each other or at the same time – if this happens, the counting starts from zero. This game is fun counting in ones or skip counting in any multiples.

- **Walk and group:** The children walk around the classroom space. When the teacher calls out a number, they get into groups of that number. Count up the groups, and how many are over. Ask the children to check your totals against the known total of children in the class. This game is also useful as a warm-up for division with remainders.

Reading numbers
Display a 3-digit number. Ask the children to read the number, identify how many hundreds, tens and ones it has, and then count on and back in steps of ones or tens from the number.

Number facts
Display a 3-digit number and play a game in which the children take turns to make up a 'fact' about a number. For example, display 355. The children might give these facts:
- 'It is 10 more than 345.'
- 'It is greater than 300.'
- 'It is between 300 and 400.'
- 'It is odd.'
- 'It is a multiple of 5.'
- 'Its tens and ones digits are the same.'

As children learn more about numbers they will be able to give more complex facts, but at the beginning of the year aim to get three or four simple facts per number.

More and fewer
Make up a set of questions based on a 3-digit number. For example, display the number 235 and ask the children to write or say the number that has:
- 5 more tens (285)
- 3 fewer ones (232)
- 2 fewer hundreds (35).

Number card games
- Which number is it? Display 3-digit numbers on cards. If you use numbers with similar digits, it makes the activity a little more challenging. Here is an example:

Say the numbers aloud in words (for example: *three hundred and twenty-five*) and ask the children to come and identify the correct number card.

- **Reverse the digits:** Let the children reverse the digits of each number on the cards and say the new numbers aloud. (Explain in advance what they should do with a zero in the ones place or do not use any numbers with zero in the ones place.)

Number meet-up

Ask the children to write down any 3-digit number. Once they have all done so, display any 3-digit number of your own. The children then have to work out the difference between your number and their number. They will need to decide whether to count back or forwards and how to make it easiest for themselves. For example, if you display 132 and the child has 265, they may count back 3 ones to get to 262, then back in tens to get to 232 and then back in hundreds to get to 132. Encourage number line jottings to support the counting at this stage. Bear in mind that some children will use larger jumps than others. Encourage discussion of different methods to help children to see the value of counting in larger groups. Make sure they understand that you get the same result – one method is simply faster (more efficient) than the other.

Target number

Give the children a target number, for example 60. Ask the children to write as many calculations as they can to make that number.

- You may want to focus on a particular operation, for example: *Write as many addition sentences as you can that equal this number.*
- Alternatively, you can leave the activity open-ended and challenge the class to find as many different operations as possible.

Show me a number

Give children sets of place-value cards (see pages 22–23). Get the children to sort their cards into ones, tens and hundreds. When they have done this, ask them to show you some numbers using only one card. For example, say: *Show me 6, 40, 500.*

Next, show the children how to put the cards together to build numbers. You may need to demonstrate. For example (putting the 10 and 5 together), say: *This is how we make 15.*

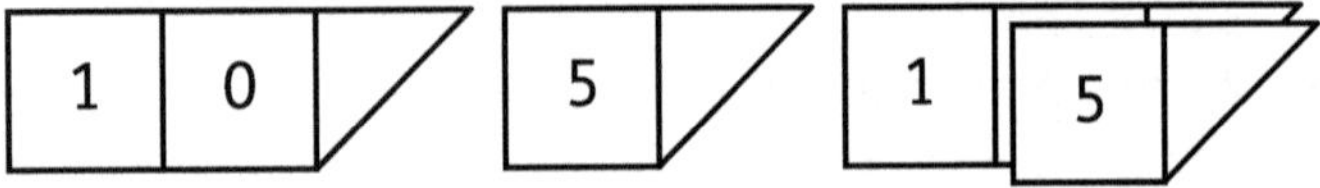

Check that the children can build different numbers with hundreds, tens and ones by calling out some numbers and having them show you. As children build numbers, they will begin to make connections and observations. They may notice that building the numbers is the same as adding numbers, for example 200 + 50 + 6 = 256. This forms the basis for partitioning numbers and written methods at later stages. Encourage the children to share their observations with the class.

Digit positions

Use place-value cards (see pages 22–23) and spend some time building numbers with the same digits in different positions, for example 139 and 931, and 217 and 712, to make sure the children understand that the position of the digit is important. Discuss the value of the digits in the numbers that the children build. Do this

by building a 3-digit number (for example, 753) using place-value cards and ask the children to say what the 7, 5 and 3 represent (700, 50 and 3).

Ask the children to make as many numbers as possible with a 1, 2 or 3 in any place. (They will need to work in groups and combine their cards to build these, or they will need to record as they make each number to keep track.) Repeat with different digits.

Place-value games

Play some games with the place-value cards (see pages 22–23) to challenge the children to listen and think carefully. For example:

- *Build a number that has digits that add up to 10. (73, 64, and so on)*
- *How many numbers less than 100 can you build with a 4 in the tens position?*
- *Build a 3-digit number with the tens digit 1 less than the hundreds digit and the ones digit 1 less than the tens digit. What is the highest number you can make? (987) What is the lowest number you can make? (210) Why?*
- *I am a number between 200 and 300 with 1 zero. What number could I be?*
- *I am a number between 400 and 600 with 1 five. What number could I be?*
- *Build a number that reads the same forwards and backwards.*

More or less

Use place-value cards (see pages 22–23) to reinforce counting activities. These activities are useful because they require the children to build numbers physically and then partition them to replace digits, which helps them make sense of calculations involving 2- and 3-digit numbers. For example:

- *Build a number that is 10 more/less than 425.*
- *Build a number that is 5 more/less than 927.*
- *Build a number that is 100 more/less than 513.*

Building numbers in place-value tables

Display a place-value table like this one and let the children copy it into their notebooks.

H	T	O

- The children pick a digit card for hundreds, a digit card for tens and a digit card for ones and place them in the place-value table to make a number (for example, 4, 5 and 7 for 475).
- The children pick three digits cards to make 3-digit number, and then make the number in the place-value table with counters or cubes.
- The children play a game where they take turns to spin a 1–6 spinner, or pick a digit card without looking. They write the number in one of the columns in the place-value table. For example, if they get 6, they may write it in the tens or ones place. Then they spin the spinner or pick a card again and write the number in the empty column. Change the aim of

the game so that sometimes the winner is the child who makes the greatest number and other times it is the child who makes the smallest number. You can include 0 as a challenge.

- A variation of the game above is to give the children a target number such as 325. The children then place their digits aiming to get a number that is as close to the target number as possible. They may not change the position of any digits as they roll subsequent numbers. The winner is the child whose number is closest to 325.
- Give the children some possible digits for each place value and ask them to work out how many 3-digit numbers they can make. Ask, for example:
- *The hundreds place can have: 2, 3, 4 or 5. How many 3-digit numbers can you make?*
- *The tens place can have: 1, 2, 3, 4, 5 or 6. How many 3-digit numbers can you make?*
- *The ones place can have: 0, 1, 2, 3. How many 3-digit numbers can you make?*

Number charts

Create number charts like the ones shown below. You could make large, laminated versions of the charts or create an electronic version.

Highlight one number of each place value on the chart.

Ask: *What number is shown on this chart?* For example, this chart shows 85:

10	20	30	40	50	60	70	80	90
1	2	3	4	5	6	7	8	9

and this chart shows 202:

100	200	300	400	500	600	700	800	900
10	20	30	40	50	60	70	80	90
1	2	3	4	5	6	7	8	9

Then ask the children to write each number in words and numerals.

Rekenrek counting and operations

Have children use rekenreks (see page 21) to model addition problems and subtraction problems.

Missing numbers

Display a number track of 3-digit numbers with some numbers missing. Point to an empty square and let the children say the number that goes in it. Repeat for each missing number. For example:

460	462		466		470

500			800		

- To make it easier, have only one or two numbers missing.
- To make it harder, leave out more numbers. Leaving out the first number means children have to work out both more than and less than a given number.

Guess the number

Play 'Guess the number' either as a class or in groups. Let the children take turns to choose a 3-digit number and write it down. The group then takes turns to ask questions to try to guess the number. The child who has the number may only answer 'yes' or 'no'.

Count in steps

Ask the children to count in given steps. Vary the steps and the number range according to what the children are working with at the moment. Start the year by revising place value to 100, then move on to numbers to 1000, for example:

- *Count from 99 to 125.*
- *Count back in twos from 250 to 240.*
- *Count in tens from 415 to 535.*
- *Count back in tens from 650 to 600.*
- *Count in hundreds from 324 to 524.*

Counting on a number line

Display a number line marked from 0 to 1000 in intervals of 10 or 100 or search online for an interactive number line.

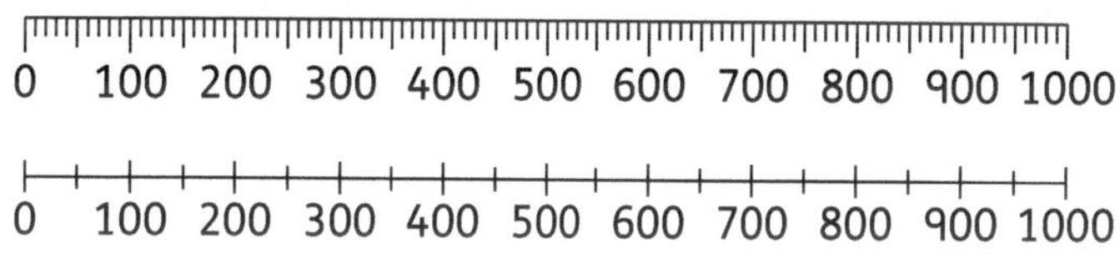

Ask questions based on counting on and counting back using a number line, for example:

- *What is 10 more than 450? (460)*
- *What is 10 less than 900? (890)*
- *What is 100 more than 600? (700)*
- *What is 100 less than 150? (50)*
- *What is 100 less than 1000? (900)*
- *What is 100 more than 350? (450)*
- *What is 100 more than 490? (590)*

Show it four ways

Give each child an 'I can show it four ways' worksheet:

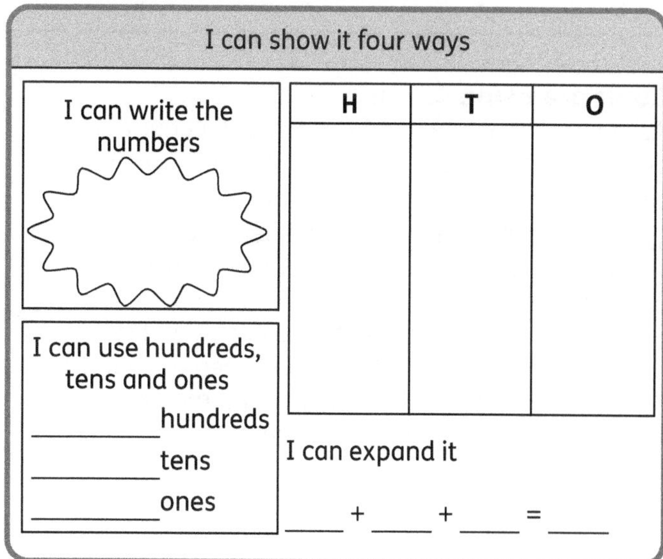

Give them a target number, for example 423, and ask them to show it in the four different ways on the worksheet.

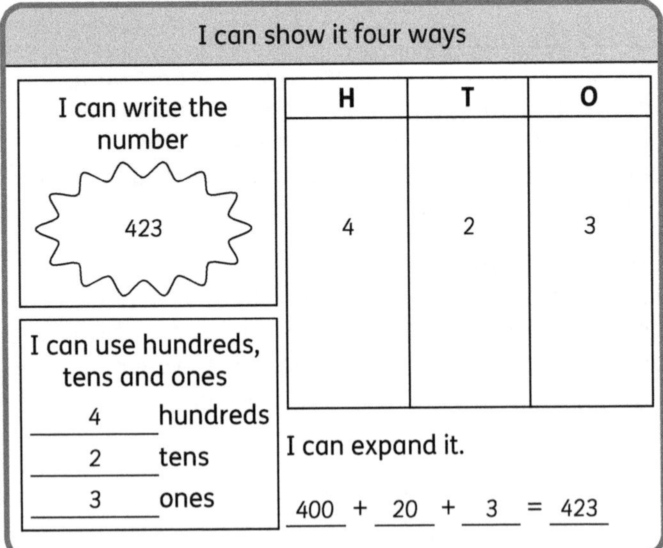

Four-part puzzles

Create puzzles that show numbers in four different ways: as a numeral (and number name), visually using base 10 flats, rods and ones, and in expanded form. Then cut these up and let the children make up each track with four equivalent representations of the number.

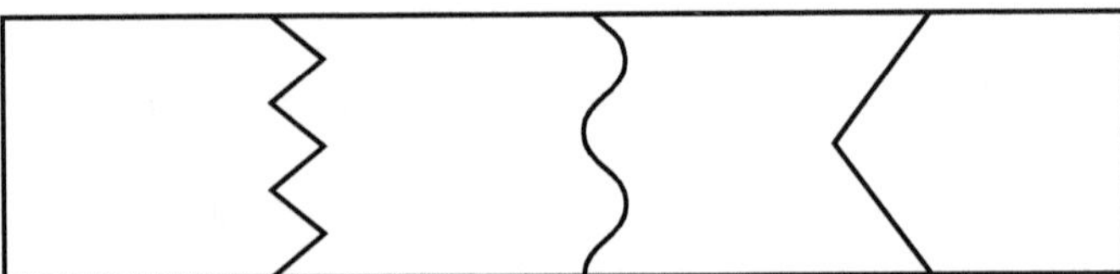

Create puzzles that show numbers in four different ways: as a numeral (and number name), visually using base 10 flats, rods and ones, and in expanded form. Then cut these up and let the children make up each track with four equivalent representations of the number.

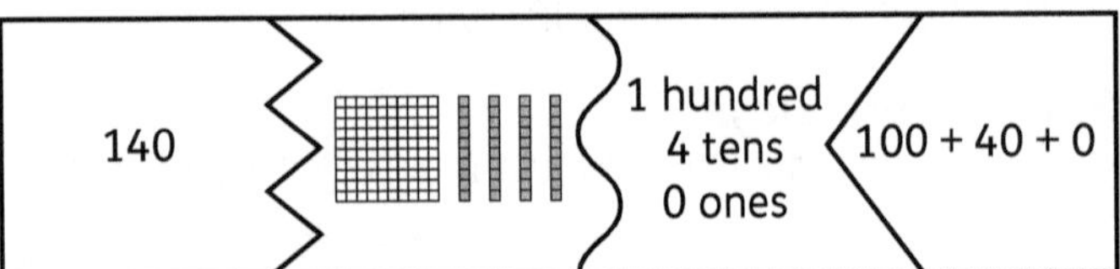

Compare using <, > or =

Ask the children to write down any 3-digit number. Write a random set of 3-digit numbers of your own on the board.

- Let the children make number sentences using your numbers and the number they have written down using the <, > or = signs.
- Let the children use mental strategies and jottings to find the sum of or difference between the numbers in their number sentences. Spend some time talking about the strategies they suggest.

Mental rounding

To practise and reinforce rounding mentally, draw a grid like this one on the board. (When practising rounding to the nearest 10, make sure none of the numbers have a zero in the ones place.)

456	275	499	109
245	501	195	108
509	824	801	876
103	562	901	105

Ask the children to copy the grid and fill it by rounding the numbers to the nearest 10 or 100. Alternatively, you can work through the grid as a class, picking a child to round each number.

Rounding challenge

Write a multiple of 100 on the board and draw six arrows pointing to it. Challenge the children to write six numbers that round to the given number to the nearest 100.

Round as you go

Ask the children to draw a three-column table like this:

100	200	300

Read out 20 numbers in the range 50 to 349. As you say each number, the children mentally round it to the nearest 100, and then write the original number in the correct column. For example, if you say 105, the children should write it in the 100 column.

Estimate

Prepare an estimation jar. Fill a glass jar or plastic container with beans or marbles. Let the children write down how many beans they think there are in the jar. They should write their estimate as a range – for example between 500 and 600.

Then show them a smaller container filled with the same objects and tell them how many there are. For example, say: *This smaller jar has 100 beans in it. Do you want to change your estimate?* Discuss how knowing the number in the smaller container helps them to estimate.

Then discuss how their estimates would change if you used larger or smaller objects to fill the jar. Discuss how you could check the estimates, perhaps by counting the objects in groups of ten or other appropriate groups.

Estimate the answer

Display a number of calculations and ask the children to estimate the answers. Make sure they focus on rounding and estimation skills – you do not want them to work out the answers. For example:

48 + 72 Estimate: 50 + 70 = 120

98 – 35 Estimate: 100 – 40 = 60

You can adapt this activity by displaying the calculations with mixed-up estimated answers for the children to match.

Estimate to make pairs

Display a set of numbers and ask the children to estimate and find pairs that meet certain conditions. For example, if you are working with multiples of 50 that make 1000, display: 45, 950, 760, 50, 450, 549, 245, 760. Ask:

- *Which pairs make a total that is close to 1000?*
- *Which pairs make a total that is more than 1000?*
- *Which make a total that is less than 1000?*

Encourage the children to verbalise their thinking, rather than just try to work out the answers. For example, they might say, '45 is close to 50, so 950 and 45 will be close to 1000.' Repeat with different combinations of numbers to fit in with classroom learning.

Do I have enough money?

Play this game with the class to practise estimating totals to 100. Say: *I have $1 and I want to buy two (melons) that each cost (60c). Do I have enough money?* You could also play this game by displaying, say, four quarters (25c coins) and asking if this is enough money to buy an item costing $1.25, and so on. Stick to low amounts at this stage. You could also use your local currency.

Encourage the children to focus on estimating and to verbalise their reasoning, rather than working out the answer.

True or false/Explain the mistake

Use 'true or false?' statements to test maths vocabulary and to provide practice in calculation skills. Discuss how they decided whether each statement is true or false. For example:

- *1 multiplied by 7 is 8.* (false)
- *2 times 3 is 6.* (true)
- *The difference between 6 and 15 is 9.* (true)
- *The sum of 3, 4 and 9 is 15.* (false)
- *15 taken away from 40 is 25.* (true)
- *Four lots of 8 are 48.* (false)
- *The product of 4 and 7 is 28.* (true)
- *50 divided by 10 is 3.* (false)
- *9 times 10 is 900.* (false)
- *45 divided by 5 is 9.* (true)

An alternative activity is 'explain the mistake'. Give the children an 'interesting mistake' and ask them to work out what mistake the person made.

Magic squares

Make sure the children understand that every row, column and diagonal in a magic square has the same total. It is easier if you tell children what that total should be.

Here are some examples that increase in difficulty.

- *Use the digits 1–9 (once each) to make totals of 15 in your magic square. Here is one possible answer.*

2	9	4
7	5	3
6	1	8

- *Now complete these magic squares so that every row, column and diagonal adds up to 15. (Give the children one or more of these examples to try, one at a time.)*

		6
3		
	2	

6		
		9
	3	

		8
7	5	

Answers:

8	1	6
3	5	7
4	9	2

6	8	1
2	4	9
7	3	5

6	1	8
7	5	3
2	9	4

Here is an example of a more challenging magic square.

- *The total of each row, column and diagonal in this magic square is 30. Complete the square.*

12		16
	18	

You could also give the children a blank square and state what total should be. The children have to find their own solutions. For example: The total of each row, column and diagonal is 34.

The children can also investigate whether adding the same number to each number in the magic square, or multiplying each number in a magic square by the same number, gives another magic square.

Bubble puzzles

Give the children a copy of this bubble diagram, or ask them to copy it:

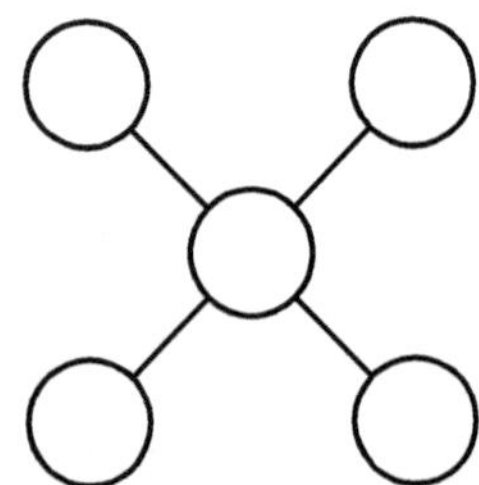

- *Write the numbers from 1, 2, 3, 4 and 5 in the bubbles so that the total along each straight line is the same.*

The children will discover, possibly by trial and improvement, that there are different ways of doing this.

Two possible solutions are:

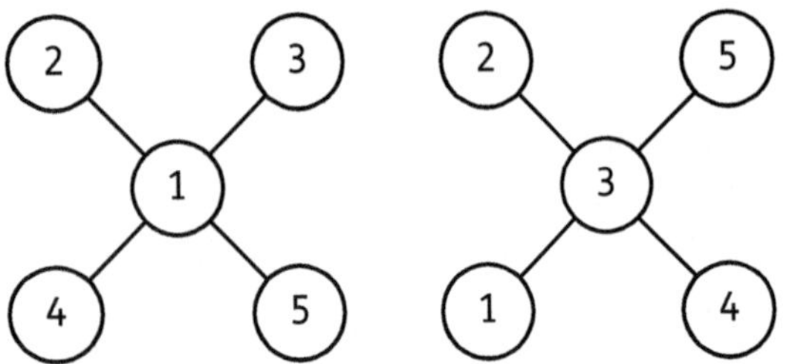

Next, give the children a diagram with seven circles (six circles around one inner circle) and the numbers 1, 2, 3, 4, 5, 6 and 7. This time the central number must be 4.

Challenge the children to complete this more challenging bubble diagram below.

- *Write all the numbers from 1 to 11 in the bubbles so that the three bubbles in each straight line have the same total. The children will need to work in pairs or groups and use trial and error.*

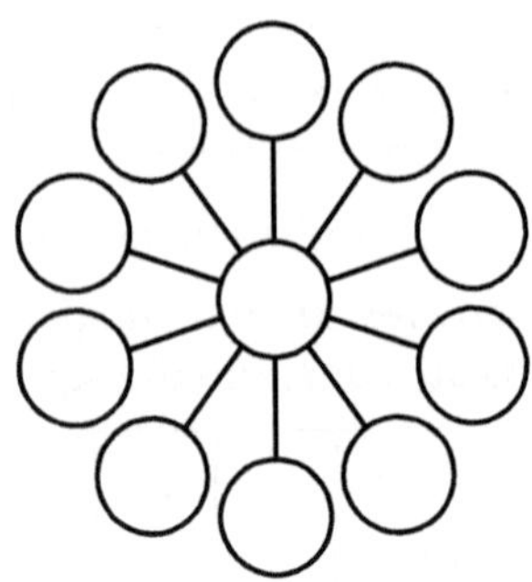

There are many possible answers. One possible answer is: 1 in the centre and, clockwise from the top, 2, 3, 4, 5, 6, 11, 10, 9, 8, 7 (all straight lines add to 14). If the children ask to be given the answer, it is worth explaining that in real life, maths is often about trying to work out a pattern or a solution that we don't already have an answer book for.

Number challenges

Give the children a set of five 2-digit numbers (for example: 43, 62, 69, 93, 87). Say:

- *Use pairs of these numbers to make (and work out) as many different subtractions as you can.*

Give the children these challenges:

- *Write four different 2-digit numbers that total 100.*
- *Write five different 2-digit numbers that total 100. Do not use 0 as a digit in any of the numbers.*

Multiplication tables

Prepare sections of multiplication tables like those shown below and display them for the children to complete. The numbers can either be in order or out of order.

×	1	4	7
2			
5			
10			

×	3	6	2
4			
2			
10			

Target games

Prepare a calculation target diagram like this one. You can write any numbers in the inner circle.

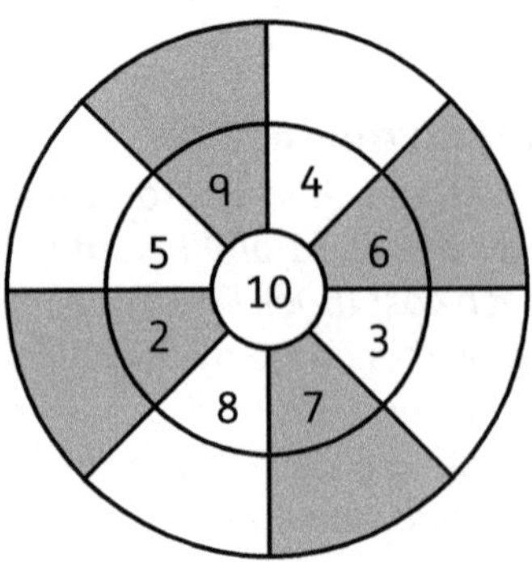

Tell the children that each sector adds up to 10 (the number in the centre) and ask them to find the missing number on the outer ring. You can adapt this by writing a small number (1–5) in the centre and then saying that the outer ring minus the inner ring will give this result. Vary the task by sometimes providing the numbers on the outer ring, sometimes the numbers on the inner ring and sometimes some of each.

You can also use target diagrams to help with multiplication. In the examples below, the children have to find the missing numbers on the diagrams:

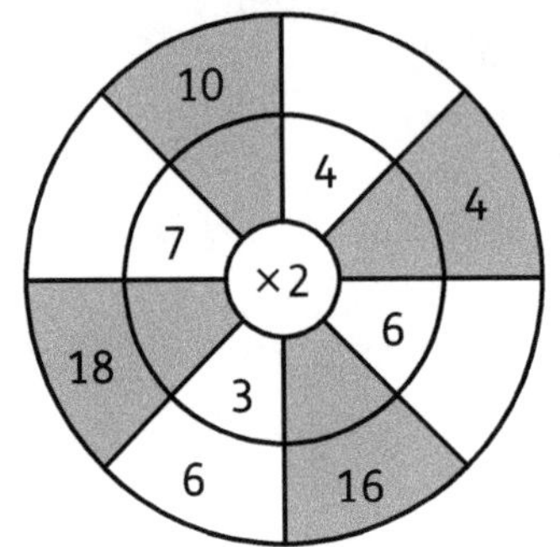

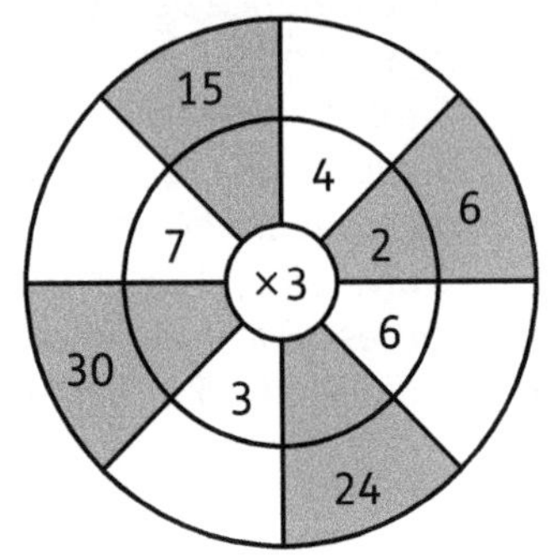

Make a total

Challenge the children to make a total of 4 using each of the digits from 1 to 9 (one at a time) and any operations they wish. They start by making 4 using only ones (1 + 1 + 1 + 1 = 4); then only twos (2 + 2 = 4 or 2 × 2 = 4); then only threes (for example, (3 × 3 + 3) ÷ 3), and so on.

A variation is to try to make all the numbers to 10 using only one number and any operations. This example uses only threes: 3 ÷ 3 = 1; (3 + 3) ÷ 3 = 2, and so on.

Cross the river game

Draw a river with several rocks in the water. Write different 2- and 3-digit numbers on each rock. Explain that the children can cross the river by jumping from one rock to the next. To do this, they must say a number sentence that gets them from their number to the number on the next rock. They can use any number operation: addition, multiplication, division or subtraction. For example, to get from 50 to 150, they could add 100 or multiply by 3.

Number sequences

Give the children a rule for generating a sequence. Demonstrate with an example, such as:

The rule is add 10 and start on 15. The first three numbers are: 15, 25, 35.

Ask the children to write down the next five numbers.

Give the rule and the starting number, and let them work out the numbers in the pattern, for example:
- *The rule is add 10. Start on 11.*
- *The rule is subtract 4. Start on 200.*
- *The rule is multiply by 2. Start on 1.*
- *The rule is half the number. Start on 520.*

You can adapt this by giving the children sequences and letting them identify the rules, for example:
- *35, 30, 25 ... (subtract 5)*
- *55, 65, 75 ... (add 10)*
- *45, 40, 35 ... (subtract 5)*
- *58, 61, 64 ... (add 3)*
- *2, 4, 8, 16 ... (double or times 2).*

Simple function machines

Prepare some function machines with only one input or output, and let the children complete them. For example:

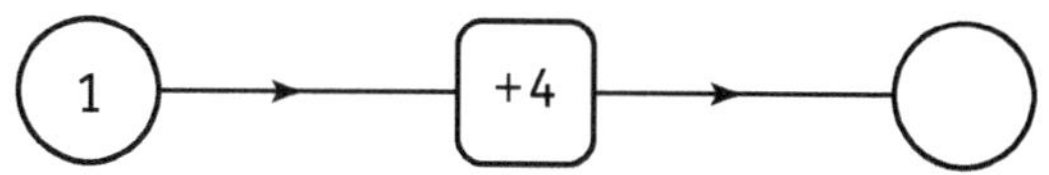

You can also use a diagram like this to show inverse operations:

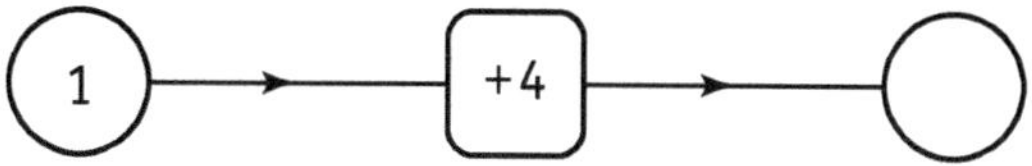

Function machines

Prepare some function machine charts and let the children complete them.

For example:

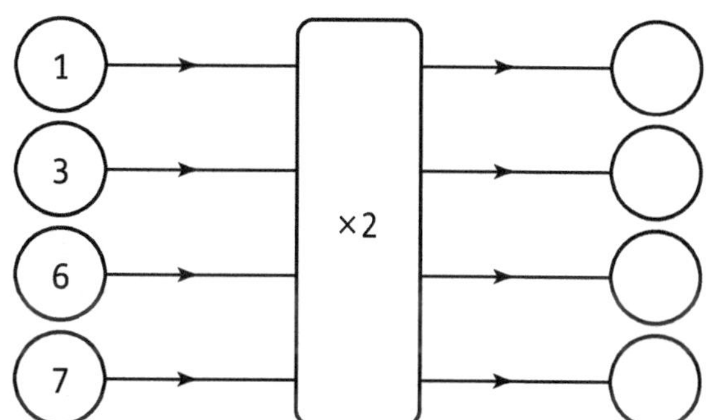

Unit 1 is different from all the other units in the course. It is a short introductory unit, which allows you to establish important norms for your class. This unit aims to prepare both the children and teachers for the work ahead. You should be able to complete this unit in one or two lessons.

As teachers of maths, it is important for us to examine and question our own ideas about who can do maths and about how mathematical ability and learning develop. From the earliest years, we need to find real, experiential ways for children to understand the following ideas:

- Asking questions is the best way to learn.
- Making mistakes is a key part of brain growth and learning.
- Everyone has the potential to learn maths. There is no such thing as a special 'maths brain'.

If we simply tell children these things, they will not believe us. We need to show them evidence of the ways in which the brain grows and learns. We also need to show them that maths can include everyone.

Learning objectives

- Establish a classroom environment conducive to thinking and working mathematically.
- Set up positive norms.
- Establish key messages of growth-mindset mathematics.

Key words

estimate   growth mindset   mindset   mistake
strategy

Unit introduction

Explain to the class that this first lesson will prepare them for learning maths this year. Write on the board: What is maths? Let the children share their many different ideas about mathematics and write them all on the board. For example, they may talk about:

- different aspects of maths, such as counting, place value, addition, measuring
- types of maths problems, such as number sentences, word problems, games and puzzles

- their feelings about maths or their expectations of maths lessons, using words such as 'hard', 'easy', 'fun', 'worry', 'scared', 'boring'.

Allow all answers. Help the children to group their ideas as you write them on the board. You could use the categories suggested in the list above, or others that emerge from the group discussion.

Seeing in different ways

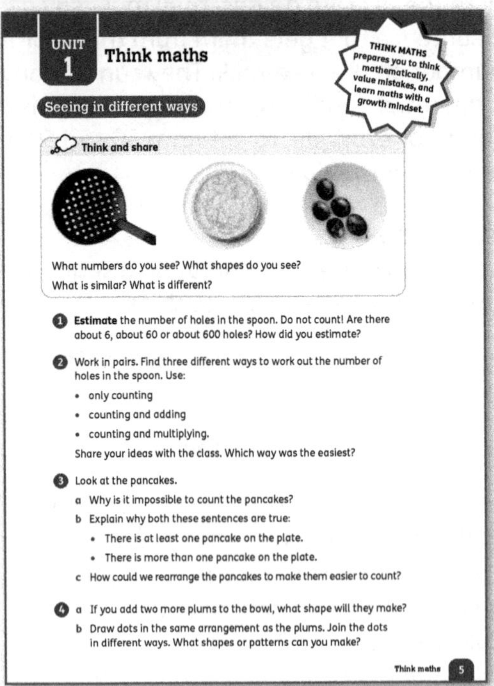

Warm-up

Revise the word *estimate* (to make a sensible guess or calculate using rounded numbers). You can use the activity 'Estimate' (page 28) as a warm-up.

Focus

- <u>Think and share</u>: Ask the children to describe what they see in the pictures on **Pupil Book 3 page 5**. Discuss question 1 with the class. Which estimate do the children find the most reasonable? Why?
- For question 2, the children need to find different ways of working out the total number of holes. Once they have discussed the different *strategies*, they can go ahead and count the holes.
- Discuss the different questions and statements about the picture of a pile of pancakes with the class.
- Talk about the arrangement of the plums. Ask what shape the children can see.

Answers for Pupil Book 3 page 5

<u>Think and share</u>: Possible answers:

Numbers: 3 items/pictures, 1 spoon, 66 holes in the spoon, 1 plate, many pancakes, 1 bowl, 5 plums

Shapes: circles, ovals and a trapezium/half a hexagon/four-sided shape; squares (the pattern of the holes in the spoon)

Similarities: circles

Differences: the numbers of circles; the number of items in each picture; the shapes the items in each picture make

1 about 60

Individual answers. For example: 'There are 9 rows and 9 columns. $9 \times 9 = 81$, so it must be less than this.' 'There are more than 6 holes in just 1 row, so the total can't be 6, and 600 would be far too many. 60 looks about right.'

2 Possible answers: Counting: count the rows horizontally, vertically or diagonally; partition the dots into shapes and count those; count in 2s or 5s.
Counting and adding: count each row or column then add the totals
Counting and multiplying: count the number of rows with different numbers of dots: 1 row each of 4 dots, 5 dots, 6 dots, 7 dots, 8 dots; 4 rows of 9 dots. Work out 4×9 and add this to the total of the other rows.

3 **a** The pancakes are arranged in a pile. We are looking from the top so we cannot see how many pancakes there are.

  b At least one means that the smallest number of pancakes there can be is one, and we can see that there is more than one.

  c Possible answers: look at the pile of pancakes from the side; lay the pancakes side by side.

4 **a** Possible answers: a circle (if we draw a curved line around the group), a flower, a hexagon (if we draw straight lines around the group)

  b Possible answers: triangles, quadrilaterals

Maths mindset

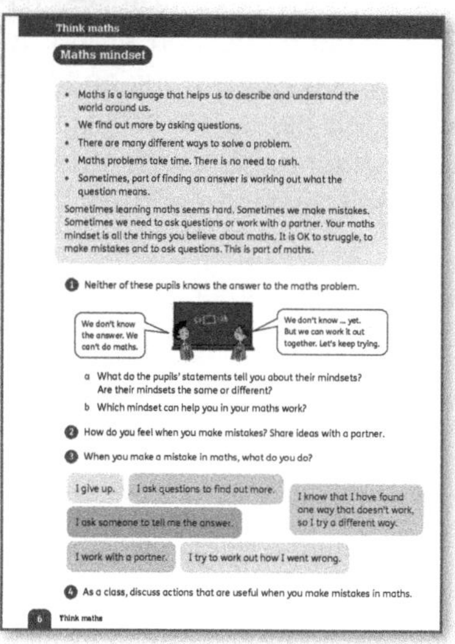

Materials

Strips of paper; small container or basket

Warm-up

- Ask the children to look again at their work from **Pupil Book 3 page 5**. Ask: *What did the questions show us about mathematics? What is mathematics?*
- Let the children share their ideas.

Focus

- Work through the information at the top of **Pupil Book 3 page 6**.
- Discuss the importance of making *mistakes* and asking questions.
- Introduce the term *mindset* – a person's attitudes and opinions, which influence how they approach challenges and new situations.
- You may also want to introduce the term *growth mindset* – the mindset of someone who knows we need to struggle and make *mistakes* in order to grow.
- Ask: *What else did you notice about the pictures and questions on page 5? What was interesting/surprising?*
- Work through and discuss question 1.
- Let the children talk in pairs about question 2 and write their ideas on strips of paper. Then collect all the children's strips of paper in a small container or basket.
- Discuss the answers that the children chose for question 3.
- Let the children suggest more strategies that could help them work out their mistakes.
- Once you have a list of strategies, tell the class you are going to read out some of the ideas from question 2. Some of the strips of paper may express positive feelings and others may express negative feelings.
- You can ask: *Does anyone else feel like this? Do these feelings help us to learn maths?* (If yes:) *Great! We know that mistakes mean we are learning and growing.* (If no:) *OK. What can we do to help when we feel like this? What can we do to understand our own mistakes?*

Answers for Pupil Book 3 page 6

1 **a** Possible answers: The first child wants to give up; he doesn't see the point of trying; he believes that if the maths is difficult, he cannot do it. The second child believes that by working together and persisting, they can learn and grow.
Their mindsets are different: the first child has a fixed mindset; the second child has a growth mindset.

  b A growth mindset can help children in their maths work.

2–**4** Individual answers

Our maths statements

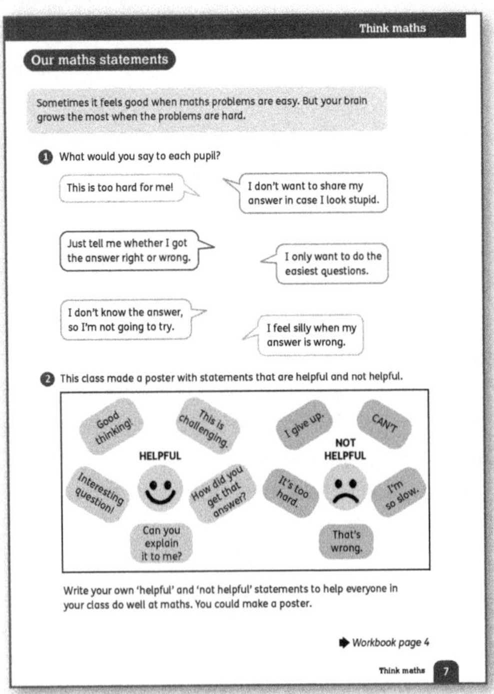

Materials
Materials for making posters

Warm-up
Allow the children to work in pairs or groups to suggest ideas for their responses to each statement in **Pupil Book 3 page 7** question 1.

Focus
- For question 2, the children suggest their own ideas for statements that are 'helpful' and 'not helpful' in maths and write these in **Workbook 3 page 4**.
- If you wish, you could collect the children's ideas to make a large class poster like the one on **Pupil Book 3 page 7**.

Answers for Pupil Book 3 page 7
1 Possible answers:
This is too hard for me! 'Try to remember how you solved a similar easier question.'

I don't want to share my answer in case I look stupid. 'Everyone makes mistakes. If you get it wrong, it will help other pupils who make the same mistake.'

Just tell me whether I got the answer right or wrong. 'It is important to know how you solved the problem so you can use the same method to solve a similar problem.'

I only want to do the easiest questions. 'When you progress in maths, you need new challenges. Working on harder questions makes you feel great!'

I don't know the answer, so I'm not going to try. 'At first, nobody knows the answer. Is there part of the question that you can try?'

I feel silly when my answer is wrong. 'Don't worry, everyone can get an answer wrong. Can you try another way to solve it?'

2 Possible answers:
Helpful: 'You did it!' 'Amazing thinking!' 'Can you tell me your way of thinking?' 'I can use counters to help me!'
Not helpful: 'I am stupid!' 'Everyone in the class can solve this except for me.' 'I won't try to solve it until the teacher helps me.'

Answers for Workbook 3 page 4
1 Possible answers: 'Good thinking!' 'Is there another way to solve this question?' 'Let's work together.'

End-of-unit check
Ask the class questions to recap the work they have covered in this unit:
- *What did you learn in this unit?* (Possible answers: 'We know that mistakes mean we are learning and growing.' 'Everyone makes mistakes.' 'It is nice to get a question right but you need new challenges.' 'Working on harder questions makes you feel great.')
- *Did anything surprise you?* (Individual answers, for example: 'It's OK to get questions wrong in maths because everyone makes mistakes and that is how we learn.')
- *Which activity did you enjoy most? Why?*
- *What do you think will help you most to succeed in maths this year?*

If you wish, you can also recap some of the concepts:
- *What is a mindset?* (a person's attitudes and opinions, which influence how they approach challenges and new situations)
- *Which kind of mindset helps us to learn mathematics?* (a growth mindset)
- *What kinds of statements help us to develop this mindset?* (Possible answers: 'Everyone makes mistakes. If you get it wrong, it will help other pupils who make the same mistake.' 'It's important to know how to solve a problem so you can use the same method to solve a similar problem.' 'When you get a question wrong, try to work out where you went wrong.')
- *Why is it important to make mistakes?* (Possible answers: 'If you get a question wrong, you can work out where you went wrong and how to get the right answer – this helps your brain grow.' 'Making mistakes and sharing them with others helps other pupils who made the same mistake.')

"

2 Number and place value

Learning objectives

- Count forwards and backwards in ones, tens and hundreds from any number between 0 and 1000.
- Identify, represent and estimate numbers using different representations.
- Understand that 10 tens = 100.
- Recognise the place value of each digit in a 3-digit number (hundreds, tens, ones).
- Understand and explain that the value of each digit is determined by its position in that number (up to 3-digit numbers).
- Compose, decompose and regroup 3-digit numbers.
- Partition 3-digit numbers to show place value.
- Order and compare 3-digit positive numbers using > and <.
- Estimate numbers of objects/people up to 1000.
- Recite, read and write numbers up to 1000 in numerals and in words.
- Round 3-digit numbers to the nearest 10 or 100.
- Solve number problems and practical problems using number facts and place value.

Key words

number names to 1000   place value   digit
hundreds   tens   ones   one thousand
place-value table   column   number sentence
difference   estimate   range   sum   partition
regular partition   irregular partition   number line
order   round   round up   round down

Unit introduction

Materials

Base-ten blocks (ones cubes, 10-rods, 100-flats) (see page 22); large copies of a 1–200 chart (1 to 10 in row 1, 11 to 20 in row 2, and so on); place-value tables (see page 23); pieces of card; counters; place-value cards (see pages 22–23); 1–9 digit cards

Teaching guidance

There are a lot of introductory activities in this unit, as it is important to use many practical activities to develop the concept of *place value* up to 1000. Choose the activities that best suit your class. This is a long unit, so you might also choose to spread out some of these activities as lesson warm-ups and consolidation work as you work through the unit.

- Work with the base-ten blocks. Ask individual children, one at a time, to each count out ones cubes. When each child reaches ten, ask them to exchange their 10 ones cubes for a 10-rod. Use questioning to build up the counting in tens gradually to 100:
  - *One, two, three, four, five, six, seven, eight, nine, ten … now we have 1 ten.* (Demonstrate exchanging ten ones cubes for a 10-rod.) *This is 1 ten.*
  - (Allow another child to count a set of ten.) *How many tens do we have now?* (2) *What number is this?* (20)
  - (Add another ten.) *Now I have 3 tens. Let's count in tens … 10, 20, 30 …* (Continue up to 100.)
  - (Demonstrate exchanging ten 10-rods for a 100). *This is one hundred. 10 tens makes …* (100)
- Use a 1–200 chart to teach the children how to say and recite numbers from 100 to 200. This will help them deal with higher numbers in the hundreds range as they work through the following activities.
- Demonstrate composing numbers on a *place-value table*. Demonstrate counting by adding blocks to the ones *column*, exchanging ones for tens every time you reach a multiple of 10. Then, demonstrate how we exchange 10 tens for 1 hundred.
- Play a 'Guess the number' game. Write some numbers on cards and stick them face down on the board. Let the children ask questions to try to work out the numbers on the cards. Explain that you will only answer 'yes' or 'no', so they need to ask questions such as: 'Is it greater than 120?', 'Is it an odd number?', 'Is it a multiple of 5?'.
- Make a large copy of the 1–200 chart for each group. Give the children a set of instructions to move counters to different positions on the grid. For example, say: *Count 10 forward, count 3 back, jump three lots of 10, go forward 50.* Later, extend this activity to revise rounding to the nearest 10. Give the children a number (for example, 143) and ask them to place their counter on the nearest 10 (140).
- Choose a start and end number with the class. Ask questions such as: *How many jumps of 10 do I need to take from the start number to the end number? How many extra jumps? Which number is halfway between these two numbers? Is the difference between the numbers more or less than 50? How can you check?* Continue with similar questions.
- Let the children play a game to make 3-digit numbers. They need three sets of 1–9 digit cards, and pick one card from each without looking. They make as many 3-digit numbers as they can with these *digits*. For example, if they pick 1, 4 and 3, they can make 134, 143, 314, 341, 413 and 431. Let the children say how many *hundreds*, *tens* and *ones* each number has, for example: '134 has 1 hundred, 3 tens and 4 ones.'

- Make a selection of 1–200 charts with some numbers missing for the children to complete. The missing numbers can be random or you can select a pattern (for example, leave out every third number, or leave out all multiples of 5 or 10).
- Cut up a 1–200 chart to make smaller sections (strips, blocks and irregular sections, such as a T-shape) and display these. Ask the children to say which numbers will appear before this section, above this section, after this section and under this section.
- Before moving on to higher numbers, provide plenty of opportunities for the children to write numerals above 100. Read or say numbers and let the children come up to the board to write the number. Focus on numbers that could be confusing, for example 154 and 145.
- Use place-value tables and place-value cards and the following activities to teach place value in 2- and 3-digit numbers: 'Show me a number', 'Digit positions', 'Place-value games', 'More or less' and 'Building numbers in place-value tables' (pages 26–27).

Revise numbers to 100

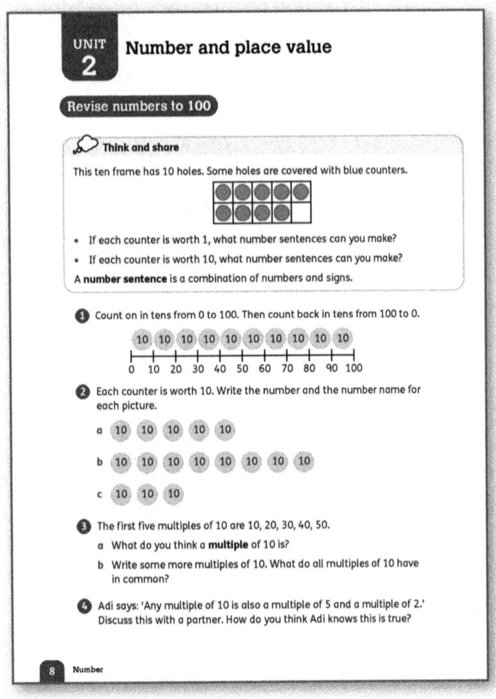

Materials
1–200 chart; 1–9 digit cards; place-value cards (see pages 22–23); counters or small stones

Warm-up
- Revise counting to 100 orally. Sitting in a circle, the children count around the group in ones, twos, fives and then tens.
- Count quickly then slowly, loudly then quietly.
- Start at different numbers (for example, 29) and let the children continue.
- Vary the activity by counting forwards and backwards.

Focus
- <u>Think and share:</u> Working in pairs, the children look at the picture at the top of **Pupil Book 3 page 8** and spend some time saying what they see. Encourage them to suggest some different ways we could express the numbers shown. Read the discussion points to the class and discuss the *number sentences* they can make.
- Use the 1–200 chart to consolidate counting on (move right and/or down) and counting back (move left and/or up). Try to elicit the pattern for counting in tens (the number one row below is ten more; the number one row above is ten less).
- In question 1 the children count on and back in tens.
- Explain that each counter is now equal to 10. To work out the totals in question 2, the children need to count in tens or add tens. You may wish to work through the first picture with them.
- Discuss the sequence 10, 20, 30, 40, 50 in question 3. Ask: *What is the pattern? How much do you add each time to get to the next number?* You can ask: *What does the word 'multiple' remind you of?* (multiply). Explain that multiples of a number are the totals when we add the number repeatedly.
- Give some examples of 2-digit numbers and have the children identify the value of each digit in the numbers in question 4. Ensure that they understand that the last digit is in the ones place and the digit to the left of it is in the tens place.

Challenge
Let the children play a game in pairs using 1–9 digit cards. One child picks two cards to make a 2-digit number. The other child picks another card to get a number of tens to count on. For example, the first child picks 3 and 4 and makes 34. The second child picks 7, and counts on 7 tens from 34.

Support
Some children may need additional support with counting up to 100 before moving on to 3-digit numbers. Use 'Place-value games' (page 26) and 'Show it four ways' (pages 27–28). Work only with tens and ones. Do not move on to 3-digit numbers until the children are more comfortable with:
- exchanging 10 ones for a set of ten
- counting from 10 to 20
- distinguishing between the *-teen* words (thirteen, fourteen, fifteen, etc.) and the multiples of 10 or *-ty* words (twenty, thirty, forty, fifty, etc.)
- representing 2-digit numbers in a place-value table
- counting on and back from any 2-digit number.

Place the children in groups. Give each group a pile of stones or counters. Let the children count the stones in different ways. For example, let them count them one at a time, or in groups of twos, threes, fives and tens.

Answers for Pupil Book 3 page 8

<u>Think and share</u>: Example answers for counters worth 1:

10 − 1 = 9 and 9 + 1 = 10

5 + 4 = 9 (adding the two rows)

5 − 4 = 1 (taking the second row away from the first)

Example answers for counters worth 10:

100 − 10 = 90 and 50 + 40 = 90

(or variations on the alternatives for counters worth 1)

1 0, 10, 20, 30, 40, 50, 60, 70, 80, 90, 100
100, 90, 80, 70, 60, 50, 40, 30, 20, 10, 0

2 a 50, fifty   b 80, eighty   c 30, thirty

3 a Possible answers: A whole number of 10s.
Any number with a zero in the ones place.
10 times any number.

b Possible answers: 60, 70, 80, 90, 100
All multiples of 10 end in zero.

4 Possible answer: 10 = 2 × 5, so any number that is a multiple of 10 is also a multiple of 2 or 5.

Understanding place value

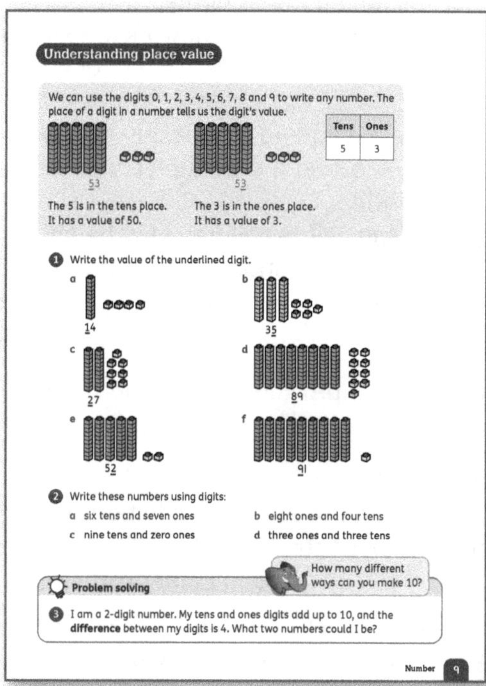

Materials

Interlocking cubes (see page 21); place-value tables (see page 23); 0–9 digit cards

Warm-up

- Represent 2-digit numbers with interlocking cubes, as shown on **Pupil Book 3 page 9**. Ask the children to identify each number and write it in a place-value table.
- Give groups of children interlocking cubes. Write a set of 2-digit numbers on the board for the children to represent using the cubes.

Focus

- Let the children work through questions 1 and 2 on **Pupil Book 3 page 9**.
- <u>Problem solving</u>: The children can use 0–9 digit cards to help them find the solutions to question 3.

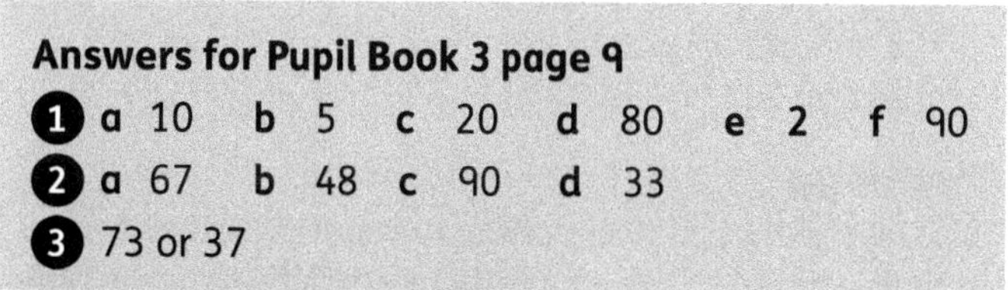

Answers for Pupil Book 3 page 9

1 a 10   b 5   c 20   d 80   e 2   f 90

2 a 67   b 48   c 90   d 33

3 73 or 37

Estimate and count

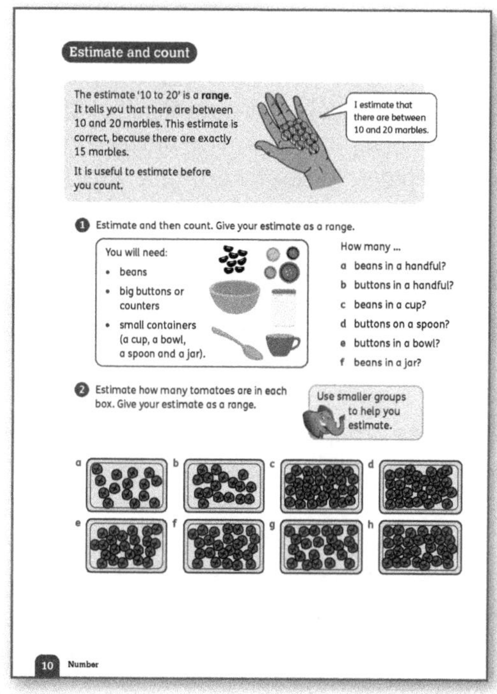

Materials

Beans, buttons and/or counters; small containers (such as cups, bowls, spoons, jars)

Warm-up

- Show a handful of small objects, such as beans or buttons. Ask the children to say how many they think you have in your hand.
- Count out the objects in fives or tens to check the children's estimate and find the exact number.

Focus

- Take another, different, handful of the same objects. Ask the children how many they think you have this time.
- Introduce the idea of a *range* by showing that each time you take a handful you get a similar amount (say, between 30 and 40 objects). If the children estimate within the range, they are likely to be close each time.
- Put some beans or other items on a sheet of paper, fairly evenly spread, and demonstrate how you can count the number of beans in a smaller group (for example, in one-quarter of the paper) and use this to estimate how many there are on the whole sheet.

- For question 1 on **Pupil Book 3 page 10** let the children work in groups to experiment, making their own handfuls and deciding on a range for their hand size. If needed, you can adapt this activity in this way: provide a selection of items on a tray and let the children estimate visually.
- The children can then work in pairs to answer question 2.

Challenge
Give the children photographs of numbers of items, for them to *estimate* 'how many' there are. Alternatively, bring patterned objects or textiles to class and let the children estimate the number of shapes, such as the number of dots on a scarf, the number of stripes on a T-shirt.

Support
For children who need more support in estimating numbers, create trays of smaller numbers of objects (under 50) for them to estimate. Give them practice in estimating first by giving three choices, for example: *Are there about 10, about 50 or about 100 objects? How did you guess?*

Interesting mistakes
Some children may make wild guesses when asked to estimate. Although we sometimes talk about estimates as 'guesses', children need to use techniques such as splitting things up into smaller groups, or looking at how much space the group of objects takes up, to work out an estimate.

> **Answers for Pupil Book 3 page 10**
> **1** Individual answers
> **2** Possible answers:
>
> | | |
> |---|---|
> | **a** between 10 and 20 | **b** between 10 and 20 |
> | **c** between 30 and 40 | **d** between 20 and 30 |
> | **e** between 20 and 30 | **f** between 20 and 30 |
> | **g** between 20 and 30 | **h** between 20 and 30 |

Tens and hundreds

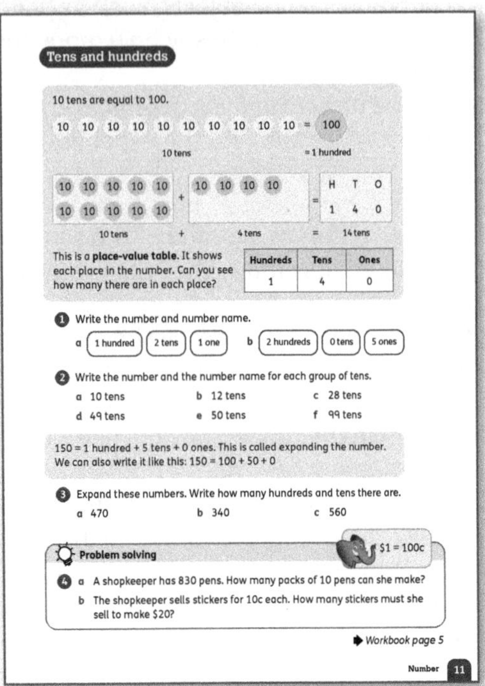

Materials
Place-value counters with 1, 10 and 100 on them and ten frames (see page 21)

Warm-up
- Use a ten frame and some place-value counters with 10 on them. Invite different children to come up and set out counters to make a given multiple of 10 (80, 50, 40, etc.). Each time, ask: *How many tens make this total?*
- Draw the children's attention to place value by composing and decomposing each number: *8 tens makes 80. The 8 shows the 8 tens and the 0 shows zero ones. How much would 1 more one make? How much would 1 less one make?* Continue with similar questions.
- Gradually move on to using more than one ten frame, so that you can demonstrate multiples of 10 greater than 100: *11 tens makes 100 and … ?* (10) Again, draw the children's attention to the place value: *11 tens is 10 tens and 1 ten. 10 tens makes 100. 100 and another ten makes 110.* Do this with more examples.

Focus
- Run through the example at the top of **Pupil Book 3 page 11** before working through the questions with the class.
- Explain question 1 and give the children time to write each number and number name.
- In question 2, the children can work independently to write the number and number name for each group of tens.
- Demonstrate expanding numbers in the two ways shown, before letting the children complete question 3 independently.
- <u>Problem solving</u>: Some children may be able to work independently on question 4. Alternatively, you can present this as a number talk (see pages 17–18).

Follow-up
Workbook 3 page 5 provides more opportunities for the children to practise working with tens.

Challenge
The children can create their own additions to make given totals, for example: 120 + 70 = 190. Ask: *How many other ways can you add tens to make an answer of 190?*

Let the children play a 'Guess my number' game in groups. Each child jots down five 3-digit numbers without letting the others see them. They then take turns to describe their numbers to a partner using clues such as as: 'It is three jumps of ten less than 150.' The partner guesses and writes down the number.

Support
Some children may need to start with a single ten frame and build up to 100. Then they can work with the range from 100 to 150, before moving up to 200, and so on. Do not rush on to greater numbers before the children have a firm grasp of these concepts:
- 10 tens make 100.
- When we add on more tens, these go into the tens place, until we have ten more – then we exchange for another 100.

Hundreds, tens and ones

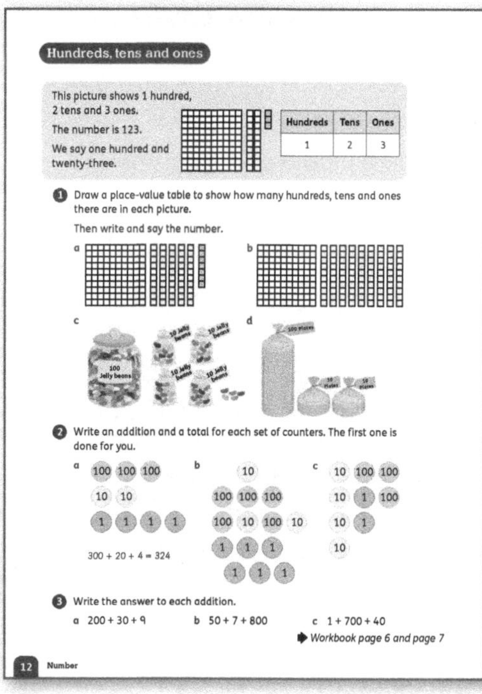

Materials
Base-ten blocks; place-value tables (see page 23);
rekenreks (see page 21); place-value cards (see
pages 22–23)

Warm-up
Use any of the following activities:
- 'Number facts' and 'Number card games' (page 25)
- 'Building numbers in place-value tables' (pages 26–27)
- 'Rekenrek counting and operations' (page 27)
- 'Show it four ways' and 'Four-part puzzles' (pages 27–28), which are useful for consolidating different representations of numbers and place value
- 'Compare using <, > or =' (page 28).

Note that you can also find rekenrek counting apps
online (see information about rekenreks on page 21).

Focus
- Work through the example at the top of **Pupil Book 3 page 12** to show the children how to draw a place-value table and use it to represent a number shown using base-ten blocks.
- Then let them work through question 1 independently.
- Discuss the collections of counters in question 2 and how to work out each number.
- Before the children complete the additions in question 3, ask them what they notice about the numbers in the three additions. (The order of the place values is different in each addition.)

Follow-up
If you have not already revised the use of place-value
cards with the children, use 'Place-value games'
(page 26), before they complete **Workbook 3 page 6**.
Workbook 3 page 7 gives more practice with place-
value tables.

Challenge
This is a creative and exploratory activity.
- Give the children counters with 1, 10 and 100 on them.
- Have them use a 3 × 3 or a 4 × 4 grid to rearrange the counters in patterns on the grid, then represent the *sum* (total) of each row, column and diagonal as a 3-digit number.
- They can also choose three colours of counters to represent 1, 10 and 100, and put these coloured counters in a 3 × 3 or 4 × 4 grid to create patterns. They can then explore what happens to the number when they change the patterns.

Interesting mistakes
The collections of counters in question 2 are deliberately
arranged in ways that can lead to confusion about
which digit comes first. We are hoping that the children
will make mistakes, in order to lead to a discussion
about the positions of the hundreds, tens and ones in a
number. For example, 4 tens, 3 hundreds and 2 ones is
342, not 432.

Answers for Workbook 3 page 6

1 1 hundred 4 tens 3 ones;
143, one hundred and forty-three
2 hundreds 3 tens 2 ones;
232, two hundred and thirty-two
5 hundreds 1 ten; 510, five hundred and ten
6 hundreds 8 tens 1 one;
681, six hundred and eighty-one

2 **a** 121 **b** 203 **c** 713 **d** 345

Answers for Workbook 3 page 7

1 Dots drawn to show the number of hundreds, tens and ones in each number.
Number sentences completed as follows:
a 117 < 137 137 > 117
b 530 > 503 503 < 530
c 927 < 972 972 > 927

2 **a** 1000 **b** 1000 **c** 1000

3 Individual answers. For example: 100 + 800
200 + 300 + 400 500 + 100 + 300

Make 3-digit numbers

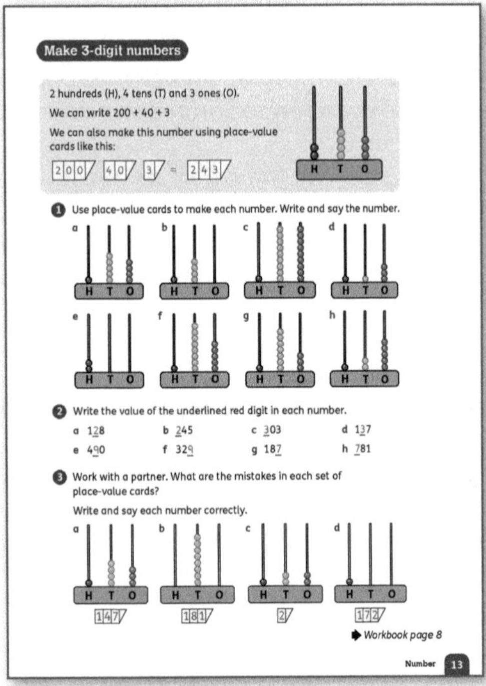

Materials
Place-value cards (see pages 22–23)

Warm-up
Use 'Place-value games' (page 26) to reinforce making 3-digit numbers.

Focus
- Discuss the example at the top of **Pupil Book 3 page 13**. Talk about how the sticks show the hundreds, tens and ones places.

- In question 1, the children use place-value cards to make each number. They may work in pairs or you could work together as a class.
- The children make the numbers using place-value cards in question 2, and then write the value of the red, underlined numbers.
- In question 3, the children work in pairs to spot the mistakes in the place-value cards and say each number correctly.

Follow-up
Workbook 3 page 8 also provides a 'spot the mistake' place-value activity.

Challenge
Use 'Guess the number' (page 27) for children who need extra challenge working with 3-digit numbers.

Support
Use 'Number meet-up' (page 26) for children who need extra support.

Answers for Pupil Book 3 page 13

1 **a** 154, one hundred and fifty-four
b 140, one hundred and forty
c 199, one hundred and ninety-nine
d 113, one hundred and thirteen
e 200, two hundred
f 185, one hundred and eighty-five
g 173, one hundred and seventy-three
h 125, one hundred and twenty-five

2 **a** 20 **b** 200 **c** 300 **d** 30
e 90 **f** 9 **g** 7 **h** 700

3 **a** The 7 should be 3.
143, one hundred and forty-three
b There should be no hundreds card and the ones card should be 0. 80, eighty
c The 100 and 20 cards are missing.
122, one hundred and twenty-two
d The 70 and 2 cards should not be there.
100, one hundred

Answers for Workbook 3 page 8

1 **a** ✗ 20 **b** ✓ **c** ✗ 800 **d** ✓
e ✗ 90 **f** ✗ 700 **g** ✓ **h** ✓
i ✗ 800 **j** ✓ **k** ✓ **l** ✗ 1000

2 Beads drawn as follows:
a 2 hundreds, 5 ones **b** 3 hundreds, 8 tens
c 4 hundreds, 7 tens

Partition numbers

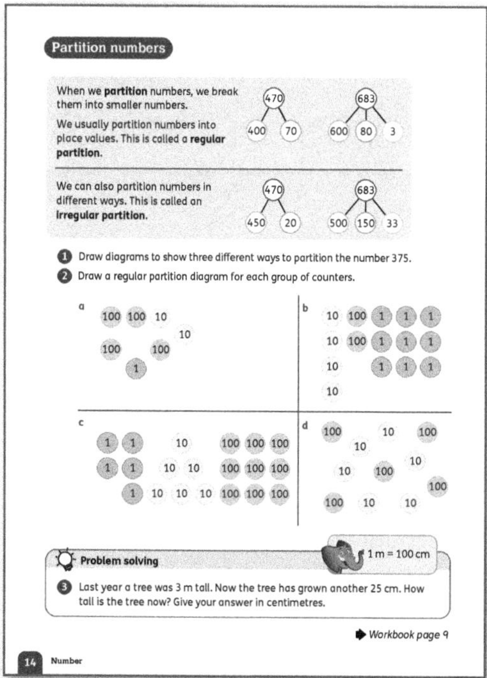

- Once the children understand partitioning, let them work independently through questions 1 and 2.
- <u>Problem solving:</u> You can offer question 3 as independent work, or as a number talk problem (see pages 17–18).

Follow-up
Workbook 3 page 9 consolidates partitioning work.

Challenge
Draw the children's attention to the way the different counters have been arranged in patterns in question 2. Let them create their own patterns using place-value counters with 1, 10 and 100, then write the numbers they have made.

Interesting mistakes
Again, the expected mistake here is that the children may mix up the positions of the place values. Making sure the digits are always in their correct places is part of the challenge of these questions. So, while it is correct to partition a number in any order (for example, 1 + 50 + 600), we must write the complete number using correct place value (651).

Answers for Pupil Book 3 page 14
1. Individual answers. For example, diagrams showing 375 partitioned as: 300, 70, 5; 200, 170, 5; 100, 250, 25
2. Diagrams showing:
   a 421 partitioned as 400, 20, 1
   b 249 partitioned as 200, 40, 9
   c 965 partitioned as 900, 60, 5
   d 560 partitioned as 500, 60
3. 325 cm

Answers for Workbook 3 page 9
1. Individual answers. For example: 100, 130, 5; 120, 15, 100
2. a 400, 70, 8   b 900, 40, 2   c 500, 30, 7
3. Individual answers
4. He has mixed up the hundreds and the tens because they are not in the usual order. Correct answer: 423
5. a 595   b 543   c 984
   d 80   e 400   f 704
   g 70   h 447   i 368

Materials
Place-value cards (see pages 22–23); base-ten blocks (see page 22); place-value counters

Warm-up
- Give the children as much practice as time allows in using place-value cards, base-ten blocks and place-value counters to represent 3-digit numbers.
- Make sure the children understand the significance of the hundreds, tens and ones by demonstrating the value of the three digits in partitioned numbers. For example, write the number 123 as 1 hundred + 2 tens + 3 ones and as 100 + 20 + 3.

Focus
- Use the examples at the top of **Pupil Book 3 page 14** to demonstrate *partitioning* in the standard way (*regular partitioning*) and non-standard ways (*irregular partitioning*). Start by drawing a structure like this on the board:

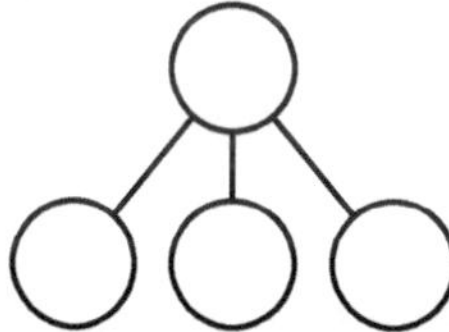

- Write a 3-digit number in the top circle, and then ask: *Who can guess what three parts we are going to break this number into?*
- The children have been working a lot with hundreds, tens and ones, so they are likely to guess. Let them make suggestions for filling in the three 'parts' that make the greater number. Then ask if anyone else can think of another way to break the number down. Draw the same structure again, and let them suggest alternatives.

Number lines

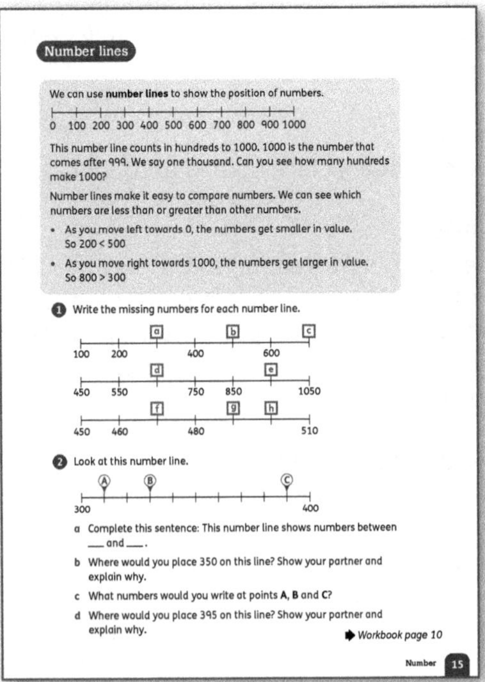

Materials
Number tracks and number lines as shown in the
Warm-up and Focus sections below

Warm-up
Prepare a set of number tracks counting back and
forwards in tens with missing numbers like these:

		235	245		265			

	400		380					

							490	500	

Have the children work out the missing numbers. They
can do this individually or you can go around the class
and let the children say the numbers.

Focus
- Remind the children that we can use a *number line*
  to count on and back from any number. Start with a
  blank line with ten intervals marked. Write a 3-digit
  number on the left-hand mark, for example 254.
  Point to the next mark and ask the class what this
  number will be (255). Point to the mark five along
  from that and ask what this number will be (260).
  Show that you can count on in ones to get there, or
  you can count in fives.
- Repeat this for several numbers and counting
  intervals, including counting back. In the early
  stages, children may count each jump. As they
  become more confident with numbers and skip
  counting, they will be able to count in larger jumps.
  The idea of 'chunking' or jumping in a larger step
  is important, as the children will use it when they
  perform calculations with larger numbers.

- Use a number line to show that we can count on
  and back in groups to add or subtract numbers.
  Demonstrate a range of different methods, for example:
  - *Counting all the numbers – for example 3 + 5 can be
    seen as 3 jumps of one then 5 jumps of one, like this:*

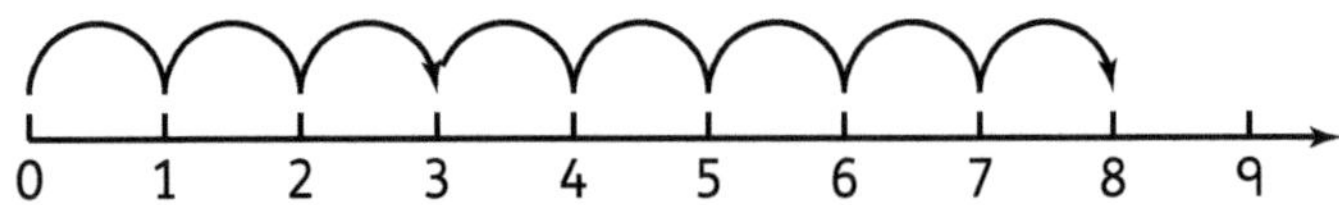

  - *Counting on – for example starting at 3 and
    counting on 5:*

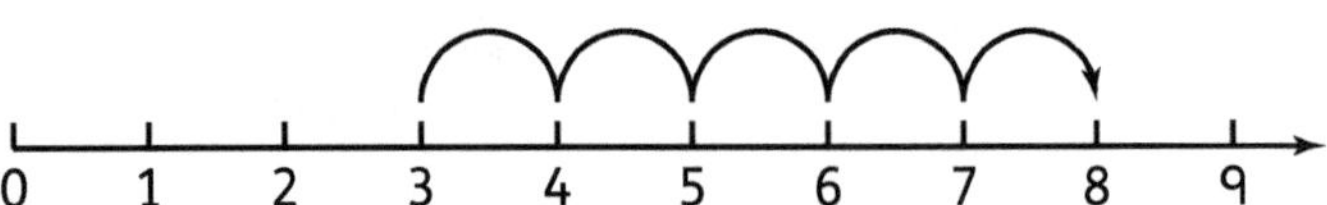

- Repeat demonstrations like this for subtraction. Make
  sure the children can read a number line properly and
  that they count the intervals and not the markers
  when they use number lines.
- Show the children that counting on and back
  in groups can also give you the answers to
  multiplication and division problems. For example:
  What is 4 lots of 3?

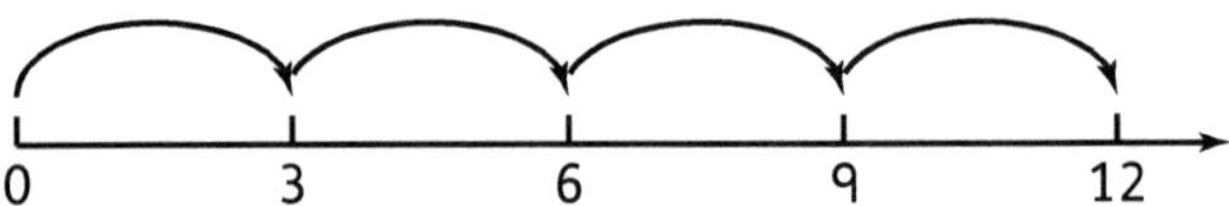

- *How many groups of 5 are there in 20?*

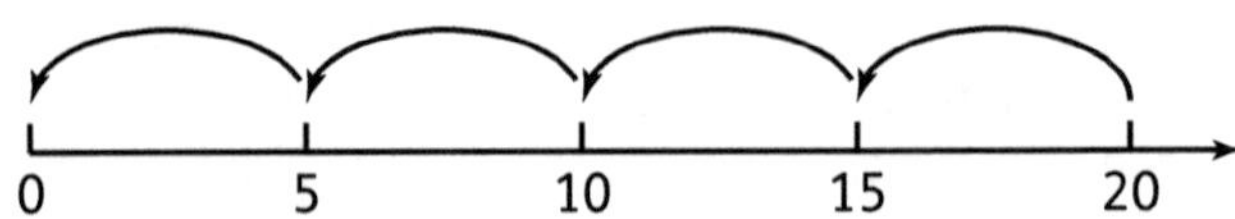

> Number lines are very powerful models for
> learning addition and subtraction, and the
> children will use these methods throughout this
> year. At later levels, they will extend their use
> of blank number lines to model more complex
> problems.

- Talk about the number line at the top of **Pupil Book 3
  page 15**. Ask: *How do you know the numbers are in
  the correct positions? What do you notice about the
  spaces between the small marks on the line? What
  do you notice about how the numbers are arranged?*
  Ensure that they understand the words *position* and
  order. Ask questions about the number line such as:
  - *Which is the lowest number on the number line?*
  - *How much does it increase with each small line?*
  - *Which number is 100 greater than (give a multiple
    of 100 between 0 and 900)?*
  - *Which number is 200 less than (give a multiple of
    100 between 200 and 1000)?*
- Count on and back in hundreds together on the
  number line.
- Spend some time talking about the concept of 1000,
  *one thousand. Ask questions such as: Where would you
  see 1000 things? Are there 1000 children in our school?
  Do you walk more or less than 1000 steps in a day?*

- 1000 is quite a large number for young children to imagine, so they need to gain a sense of its magnitude. A useful visual technique is to make piles of ten counters and say, *Let's make each counter equal to one dollar* (or substitute the local currency unit). Count out towers of ten, and draw some $10 notes. Then draw a $100 note and ask how many $10 notes we need to make $100. Repeat with ten $100 notes to make $1000.
- Explain that in question 1 on **Pupil Book 3 page 15** each letter represents a missing number on the number lines. The children work out the interval between the numbers, and use this to work out the missing numbers. Note that the three number lines each have their own range and intervals.
- Ask the children to look at each number line and say what they see. Ask:
  - *Which numbers are on the number line?*
  - *How can we work out the missing numbers?*
  - *What clues do we have?*
- Give the children time to discuss question 2 and work out that the range 300 to 400 is divided into ten equal parts. We know that 10 tens makes 100, so each interval is worth ten.

Follow-up
Use the questions on **Workbook 3 page 10** to consolidate the work on number lines.

Support
Use 'Counting on a number line' (page 27) to provide additional support.

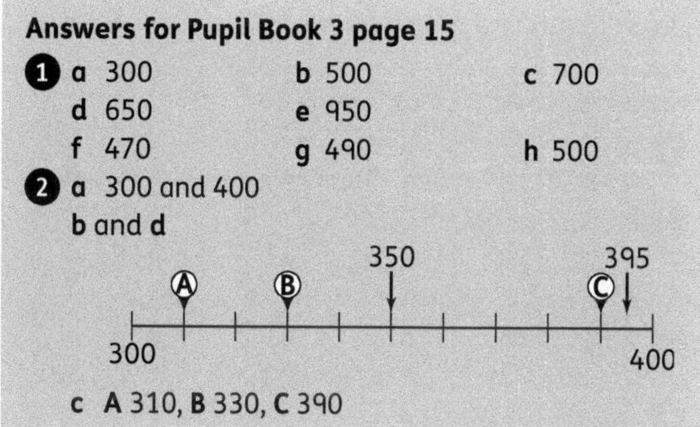

Answers for Pupil Book 3 page 15

1
a 300 b 500 c 700
d 650 e 950
f 470 g 490 h 500

2
a 300 and 400
b and d

		350		395
(A)	(B)	↓		(C)↓
300				400

c A 310, B 330, C 390

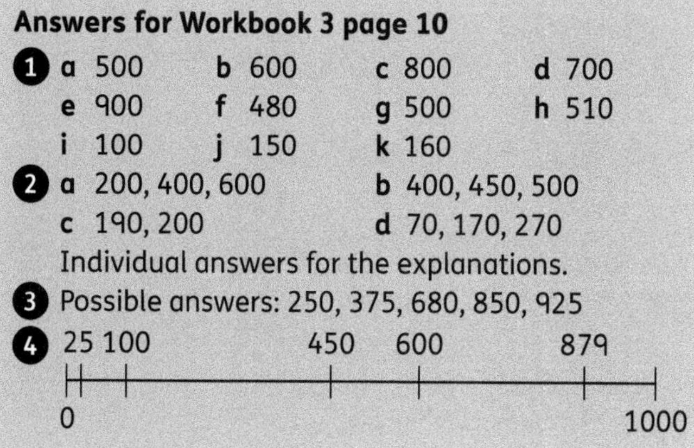

Answers for Workbook 3 page 10

1
a 500 b 600 c 800 d 700
e 900 f 480 g 500 h 510
i 100 j 150 k 160

2
a 200, 400, 600 b 400, 450, 500
c 190, 200 d 70, 170, 270
Individual answers for the explanations.

3 Possible answers: 250, 375, 680, 850, 925

4

25	100		450	600		879	
0							1000

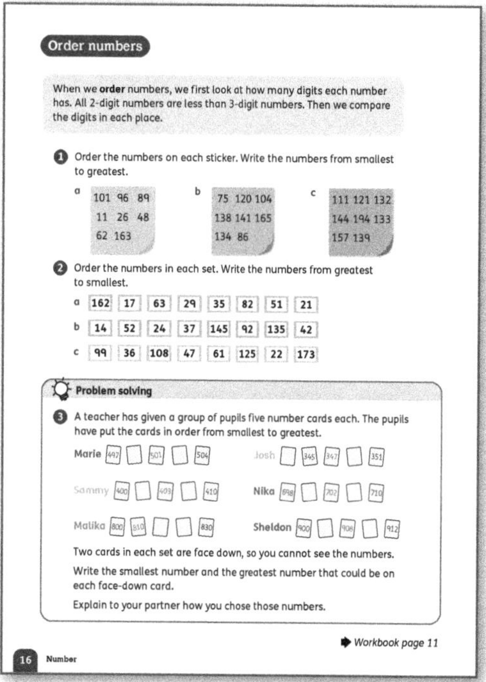

Order numbers

When we **order** numbers, we first look at how many digits each number has. All 2-digit numbers are less than 3-digit numbers. Then we compare the digits in each place.

1 Order the numbers on each sticker. Write the numbers from smallest to greatest.

a 101 96 89 11 26 48 62 163
b 75 120 104 138 141 165 134 86
c 111 121 132 144 194 133 157 139

2 Order the numbers in each set. Write the numbers from greatest to smallest.

a 162 17 63 29 35 82 51 21
b 14 52 24 37 145 92 135 42
c 99 36 108 47 61 125 22 173

Problem solving

3 A teacher has given a group of pupils five number cards each. The pupils have put the cards in order from smallest to greatest.

Marie [442] [ ] [501] [ ] [504]   Josh [ ] [345] [347] [ ] [351]
Sammy [400] [ ] [409] [ ] [410]   Nika [896] [ ] [902] [ ] [910]
Malika [800] [610] [ ] [ ] [830]   Sheldon [900] [ ] [906] [ ] [912]

Two cards in each set are face down, so you cannot see the numbers.
Write the smallest number and the greatest number that could be on each face-down card.
Explain to your partner how you chose those numbers.

➡ Workbook page 11

16 Number

Materials
Number cards with a 2- or 3-digit number on each; number lines

Warm-up
- Give five children a number card each showing a 2- or 3-digit number. Ask the children to arrange themselves in order. The other children can give them tips on how to do this and can check the order.
- Repeat with different number cards and different children.

Focus
- Read through the information at the top of **Pupil Book 3 page 16** with the children. Show a 2-digit and a 3-digit number. Ask: *How do we know that this 3-digit number is greater than this 2-digit number?* (The 3-digit number has some hundreds and the 2-digit number does not.)
- Ask the children to work through questions 1 and 2. Allow them to use number lines to order the numbers if they need to.
- <u>Problem solving:</u> For question 3, the children can discuss in pairs how to work out the largest and smallest possible numbers for each card.
- For blank cards between two given numbers, the smallest number is one greater than the number to the left and the greatest number is one less than the number to the right. Where there are two blank cards side by side, the children need to consider that both numbers need to fit within the range.

Follow-up
Use **Workbook 3 page 11** to reinforce and consolidate the skill of *ordering* numbers.

Compare numbers

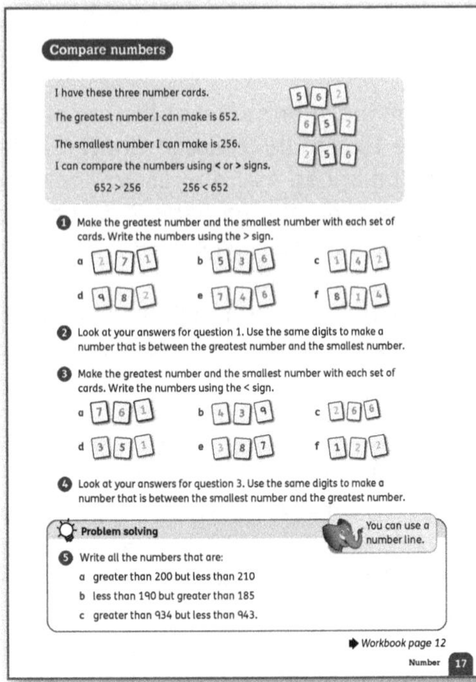

Materials
1–9 digit cards; 1–100 number line or 1–100 chart

Warm-up
- Give each group of children three 1–9 digit cards and ask them to make the smallest number they can using all three cards. They should write the number down and then make the greatest number.
- Revise the correct use of the < and > symbols with the class. Ask them to write two statements using the two numbers they made, one with < and one with >.

Focus
- Ask the children to work through questions 1–4 on **Pupil Book 3 page 17**.
- In questions 2 and 4, there are several possible numbers for each question (except questions 4c and 4f). Let the children explore how many different numbers they can find.
- Problem solving: The children could use a 1–100 number line or 1–100 chart to help them find all the possible answers to question 5.

Follow-up
Let the children work independently to complete **Workbook 3 page 12** and have them check each other's answers and discuss any differences that arise.

Support
Children frequently have difficulty remembering which of the signs < and > means 'more than' and which means 'less than'. Show the children that the < looks a bit like an L, so it might help them to remember that < means 'less than' if they think that '<ess' spells 'less'.

Round to the nearest 10

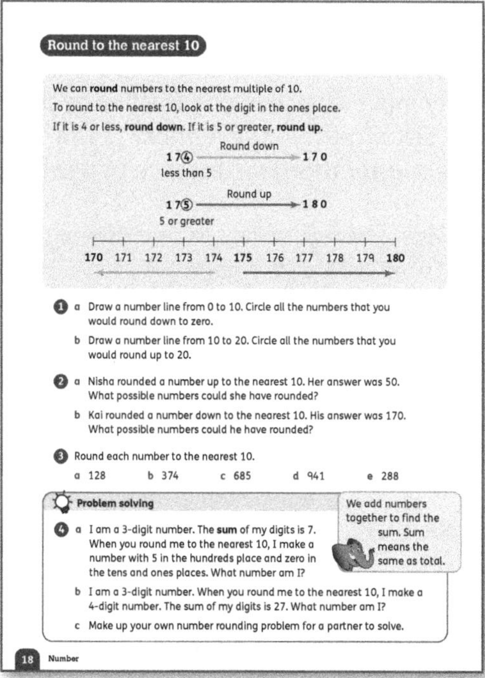

Round to the nearest 10

We can **round** numbers to the nearest multiple of 10.

To round to the nearest 10, look at the digit in the ones place.

If it is 4 or less, **round down**. If it is 5 or greater, **round up**.

Round down
17④ ⟶ 170
less than 5

Round up
17⑤ ⟶ 180
5 or greater

170　171　172　173　174　**175**　176　177　178　179　**180**

1　a　Draw a number line from 0 to 10. Circle all the numbers that you would round down to zero.

　　b　Draw a number line from 10 to 20. Circle all the numbers that you would round up to 20.

2　a　Nisha rounded a number up to the nearest 10. Her answer was 50. What possible numbers could she have rounded?

　　b　Kai rounded a number down to the nearest 10. His answer was 170. What possible numbers could he have rounded?

3　Round each number to the nearest 10.

　　a　128　　b　374　　c　685　　d　941　　e　288

Problem solving

4　a　I am a 3-digit number. The **sum** of my digits is 7. When you round me to the nearest 10, I make a number with 5 in the hundreds place and zero in the tens and ones places. What number am I?

　　b　I am a 3-digit number. When you round me to the nearest 10, I make a 4-digit number. The sum of my digits is 27. What number am I?

　　c　Make up your own number rounding problem for a partner to solve.

> We add numbers together to find the sum. Sum means the same as total.

18　Number

Materials

Number tracks (see Warm-up below); number lines; ten frames (see page 21) and counters

Warm-up

- Prepare several 'rounding to the nearest 10' number tracks like this one:

40	41	42	43	44	45	46	47	48	49	50

- Use this number track to show the children how to *round* numbers. Show them that the numbers 41, 42, 43 and 44 are closer to 40 than to 50. These numbers *round down* to 40. Similarly, 46, 47, 48 and 49 are closer to 50 than to 40. These numbers *round up* to 50.
- Point out that 45 is in the middle. Remind them that mathematics relies on rules and that, all over the world, the rule is that if the number ends in 5, it rounds up to the next place. So, 45 rounds up to 50.
- Point to various numbers on different number tracks and ask the children to say which multiple of 10 the number rounds to. If you like, use some vertical number tracks too.

Focus

- Work through the examples at the top of **Pupil Book 3 page 18** with the class.
- Then draw a number line from 0 to 10. Pick numbers at random from the number line for the children to round up or down.
- Change the number line to any set of 3-digit numbers between two multiples of 10, for example from 240 to 250.
- Work through questions 1–3 with the class, or have the children work through them independently, depending on what is suitable for your class.
- <u>Problem solving</u>: In question 4, the children solve number riddles. They can work in pairs, and then compare strategies with the class.

Support

If any children are struggling with the idea of rounding to the nearest 10, use visual aids such as number lines or ten frames. You can set out several ten frames and use counters to represent a number such as 31, then 32, then 33 … adding a counter each time and asking: *Is it closer to 30 or 40?* This example shows 37:

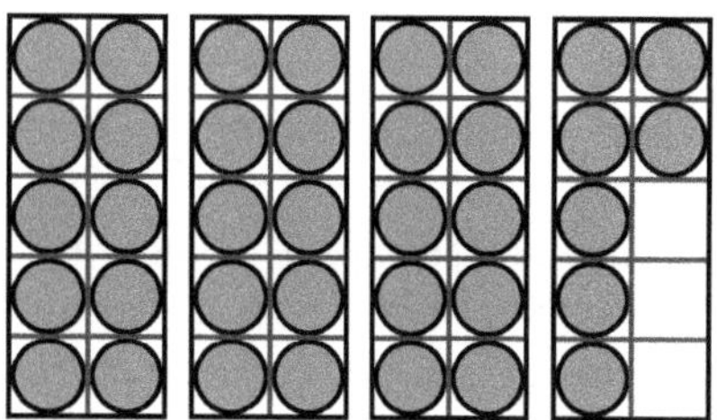

When we build up tens visually, it is easy to see that when more than half (five) of the spaces are full, we are closer to the next ten. Remind the children that when a frame is half full (5 counters) we always round up.

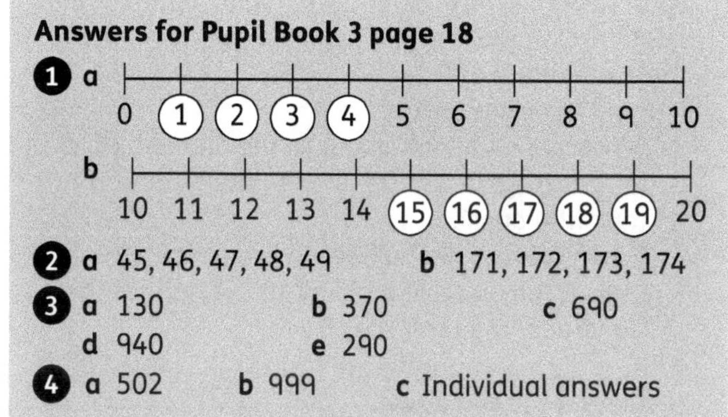

Answers for Pupil Book 3 page 18

1　a　Number line 0 to 10 with 1, 2, 3, 4 circled.

　　b　Number line 10 to 20 with 15, 16, 17, 18, 19 circled.

2　a　45, 46, 47, 48, 49　　b　171, 172, 173, 174

3　a　130　　b　370　　c　690
　　d　940　　e　290

4　a　502　　b　999　　c　Individual answers

Round to the nearest 100

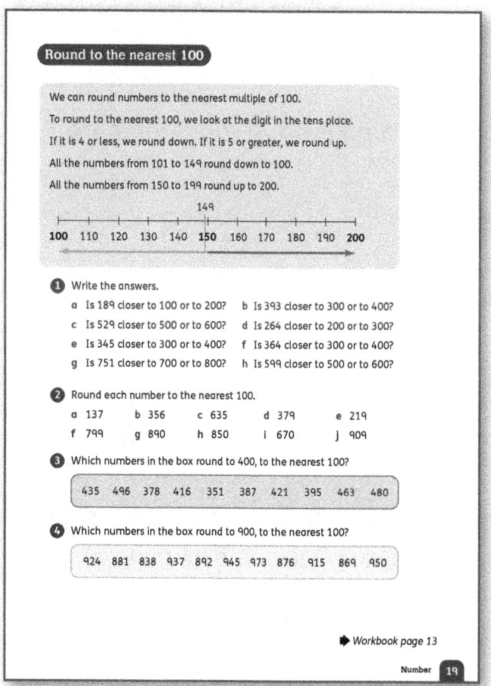

Round to the nearest 100

We can round numbers to the nearest multiple of 100.

To round to the nearest 100, we look at the digit in the tens place.

If it is 4 or less, we round down. If it is 5 or greater, we round up.

All the numbers from 101 to 149 round down to 100.

All the numbers from 150 to 199 round up to 200.

149

100　110　120　130　140　**150**　160　170　180　190　**200**

1　Write the answers.

　　a　Is 189 closer to 100 or to 200?　　b　Is 393 closer to 300 or to 400?

　　c　Is 529 closer to 500 or to 600?　　d　Is 264 closer to 200 or to 300?

　　e　Is 345 closer to 300 or to 400?　　f　Is 364 closer to 300 or to 400?

　　g　Is 751 closer to 700 or to 800?　　h　Is 599 closer to 500 or to 600?

2　Round each number to the nearest 100.

　　a　137　　b　356　　c　635　　d　379　　e　219
　　f　799　　g　890　　h　850　　i　670　　j　909

3　Which numbers in the box round to 400, to the nearest 100?

　　435　496　378　416　351　387　421　395　463　480

4　Which numbers in the box round to 900, to the nearest 100?

　　924　881　838　937　892　945　973　876　915　869　950

➜ Workbook page 13

Number　19

Materials

Sticky notes and a large number track as shown in the Warm-up below; small objects such as stones or beads

Warm-up

- Ask the children to give some examples of numbers between 400 and 500. Jot the examples on the board or on sticky notes.
- Draw a number line or track marked in intervals of ten, like this one:

400	410	420	430	440	450	460	470	480	490	500

- Call out one of the numbers that the children suggested. Ask individual children to come up to the board. They either place the sticky note with that number above the number it is closest to on the number line or track or draw an arrow from the number to its position.
- For each number, ask: *Is it closer to 400 or 500?* For example, 421 will be positioned above 420, and it will be clear that it is closer to 400 than 500. Explain that when we work with big numbers, we sometimes round them to the nearest 100. For example, if there are 417 people at an event, we might say there are about 400. If there are 489 people, we might say there are about 500.
- Remind the class that:
  - When we round a number to the nearest 10, we look at the place of the digit to the right of the tens (the ones).
  - When we round a number to the nearest 100, we look at the place of the digit to the right of the hundreds (the tens).
- If the digit is 5 or greater we round up. If the digit is 4 or less we round down.

Focus

Work through the examples at the top of **Pupil Book 3 page 19** before the children work independently through questions 1–4.

Follow-up

Use **Workbook 3 page 13** to consolidate work on rounding.

Challenge

Ask these questions:

1 *What are the smallest number and the largest number that round to each of these numbers to the nearest 100?*

  a 200 (150, 249)   **b** 300 (250, 349)

  c 400 (350, 449)   **d** 700 (650, 749)

2 *What do all the answers to question 1 have in common?* (Smallest end in 50, largest end in 49)

3 *What are the possible numbers that round to 250 to the nearest 10 but round to 300 to the nearest 100?* (251, 252, 253, 254)

Support

The children may struggle to round numbers to a given place if their understanding of place value is shaky.

- Reinforce the 'nearest 10' and 'nearest 100' using number lines.

- If the children struggle to read the position of numbers when the number line is divided into intervals other than ones, spend some time showing them how to read the labels and work out the intervals first.
- Physically move objects along the line to show them that the distance from the number to the closest 100 (or 10) is shorter than the distance to the other 100 (or 10).

Answers for Pupil Book 3 page 19

1
a 200	**b** 400	**c** 500	**d** 300
e 300	**f** 400	**g** 800	**h** 600

2
a 100	**b** 400	**c** 600	**d** 400
e 200	**f** 800	**g** 900	**h** 900
i 700	**j** 900		

3 435, 378, 416, 351, 387, 421, 395

4 924, 881, 937, 892, 945, 876, 915, 869

Answers for Workbook 3 page 13

1
a 198	**b** 328	**c** 106
d 475	**e** 398	**f** 260
g 50	**h** 301	**i** 530

End-of-unit check

Ask questions to assess the children's understanding of counting, place value and rounding. Ask, for example:

- *How many tens and how many ones are there in the number 29?* (2 tens, 9 ones) *What about 92?* (9 tens, 2 ones)
- *A number has 2 hundreds, 3 tens and 6 ones. What is the number?* (236)
- *What number is missing from the sequence 172, 173, ___, 175?* (174)
- *What number appears immediately above 173 on a 1–200 chart?* (163) Continue with similar questions.
- *What number appears immediately below 145 on a 1–200 chart?* (155) Continue with similar questions.
- *How many tens do you need to make a total of 200?* (20 tens)
- Give the children a 3-digit number. *Can you make this number with place-value cards? Can you say the number?*
- *What is the value of the '2' in the number 235?* (200)
- *What is the value of the '5' in the number 357?* (50)
- *What is the value of each digit in the number 372?* (The 3 is worth 300, the 7 is worth 70, the 2 is worth 2.)
- *Complete this number pattern: 610, 615, 620, ___, ___, ___, ___, ___?* (... 625, 630, 635, 640, 645)
- Give a 2-digit number. *Round this number to the nearest ten.*
- Give a 3-digit number. *Round this number to the nearest hundred. Round it to the nearest ten.*

Learning objectives
- Measure, compare, add and subtract lengths in m, cm and mm.
- Use instruments that measure length in m, cm, mm.
- Estimate and measure length in cm, m, km.
- Understand the relationship between units.
- Round up and down to estimate and measure.

Key words
length   width   height   centimetre   metre
millimetre   estimate   measure   round

Unit introduction

Materials
Metre rules; string and scissors; 10-centimetre strips of paper marked in centimetres; rulers; strips of paper and sticky tape

Teaching guidance
Here are some ideas for practical measuring activities:
- Reintroduce the children to the *metre* as a unit of length and show them a metre rule. Give each child or group of children a metre rule and get them to *measure* the *lengths, heights* and *widths* of various objects in the classroom. The children can refer to the lengths they measure as 'less than 1 m', 'between 1 m and 2 m' or 'more than 2 m'.
- Give each child a piece of string 1 m long, which they should fold in half and cut into two equal pieces. Explain that each piece of string is now half a metre long. Ask the children to measure the objects from the first activity again, this time to the nearest metre and half metre.
- Challenge the children to say why it might be difficult to measure lines that are not straight. Ask them how we could measure. Show them how we can place a piece of string along a curved line and then hold it straight next to a metre rule to measure it.
- Revise measurement in *centimetres* by discussing how to use a 10-cm strip or a ruler to measure accurately. Point out the scale and explain that we measure by positioning the 0 at one end of the object we are measuring and reading the scale at the other end. If you are using rulers, instruct the children to ignore the *millimetre* gradations between centimetre marks for the time being. All measurements should be to the nearest centimetre.
- The children could use a ruler to measure the length of classroom objects in centimetres. The children could draw polygons and measure the length of each of the sides in centimetres.

- The children could join strips of paper together to make a strip 2 m long, and use a ruler to calibrate the strip in centimetres. The children can then use this to measure their heights.

Measuring length

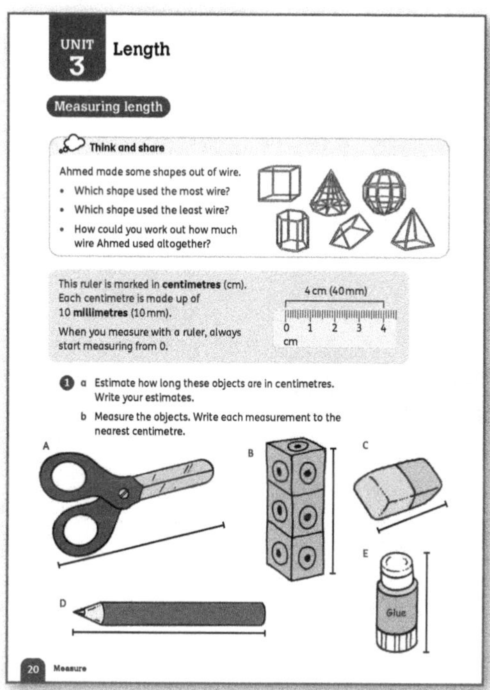

Materials
Rulers

Warm-up
<u>Think and share:</u> Begin by discussing the questions about the wire shapes on **Pupil Book 3 page 20**. Let the children work in pairs and groups to talk about what they see in the picture. They can discuss the questions and give reasons for their answers.

Focus
- Revise how to use a ruler. Although the children can see centimetres and millimetres marked on the ruler, at this stage they only need to measure in centimetres.
- Demonstrate the correct technique for measuring a line with a ruler, ensuring that the 0 mark lines up with one end of the line.
- You can work through question 1 orally or let the children *estimate* and then measure the objects independently.

Support
Some children may find it very difficult to make realistic estimates of length. Suggest that they use different objects as reference to help when they are estimating. For example, the height of their book is about 20 cm, so five books laid end to end measure about 1 m.

When working with smaller lengths, they can measure small objects (such as a pencil sharpener, a paper clip, the length of their thumb) and use these for reference when estimating other lengths.

> **Answers for Pupil Book 3 page 20**
>
> <u>Think and share:</u> The sphere used the most wire.
>
> The square-based pyramid used the least wire.
>
> We can work out how much wire Ahmed used altogether by measuring all the edges and adding them together.
>
> We can do this more quickly by counting the number of edges of the same length. Then we can measure one of each of these sides. Then we multiply each length by the number of edges with this length, and add all the totals together.
>
> **1** Measurements to the nearest centimetre:
>
> | | | |
> |---|---|---|
> | **A** 7 cm | **B** 5 cm | **C** 3 cm |
> | **D** 8 cm | **E** 4 cm | |

Measure paths

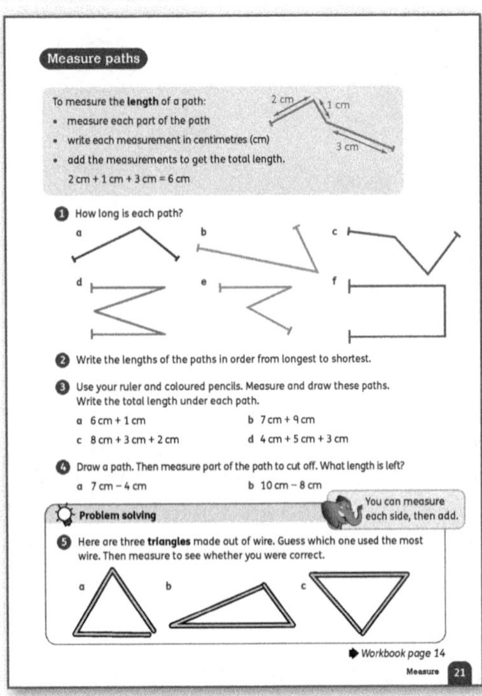

Materials

Rulers

Warm-up

- Draw a path made up of straight lines on the board, similar to those shown on **Pupil Book 3 page 21**. Tell the children that this is the path that an ant follows.
- Demonstrate how to use a ruler to draw a line 3 cm long, then turn the ruler to draw another line 2 cm long roughly at right angles to the first line.
- Ask the children to draw their own path for an ant using a ruler and pencil. Each section of the path should be a whole number of centimetres long.

Focus

- Show the children how to measure compound lengths like the paths on **Pupil Book 3 page 21** by measuring each straight line section and adding to find the total. This skill is important for later work on perimeter and area.
- For questions 1 and 2, let the children work in pairs to measure, check and calculate each length.
- Let the children complete the drawing activities in questions 3 and 4 independently.
- <u>Problem solving:</u> The children can work independently on question 5.

Follow-up

Consolidate and assess this lesson by asking the children to complete **Workbook 3 page 14**.

Challenge

Give the children this activity:

- *I have a piece of wire 28 cm long.*
- *Draw sketches of different polygons (closed shapes with straight sides) that you could make.*
- *Each polygon must use the whole piece of wire.*
- *Try to make a square and some different rectangles and triangles. What other shapes can you make?*

> **Answers for Pupil Book 3 page 21**
>
> **1**
> | | | |
> |---|---|---|
> | **a** 5 cm | **b** 7 cm | **c** 6 cm |
> | **d** 12 cm | **e** 7 cm | **f** 10 cm |
>
> **2** 5 cm, 6 cm, 7 cm, 7 cm, 10 cm, 12 cm
>
> **3** Accurate drawings of paths made up of the following sections:
>
>   **a** 6 cm and 1 cm. Total length: 7 cm
>   **b** 7 cm and 9 cm. Total length: 16 cm
>   **c** 8 cm, 3 cm and 2 cm. Total length: 13 cm
>   **d** 4 cm, 5 cm and 3 cm. Total length: 12 cm
>
> **4** **a** 3 cm   **b** 2 cm
>
> **5** The side lengths and total perimeter of each triangle are:
>
>   **a** 3 cm + 3 cm + 3 cm = 9 cm
>   **b** 5 cm + 4 cm + 2 cm = 11 cm
>   **c** 4 cm + 3 cm + 3 cm = 10 cm
>   Triangle **b** used the most wire

> **Answers for Workbook 3 page 14**
>
> **1** **a** 1 + 2 + 3 + 2 + 1 + 3 = 12 cm
>   **b** 3 + 3 + 1 + 5 + 3 = 15 cm
>   **c** 4 + 1 + 4 + 2 + 3 + 3 + 2 = 19 cm
>
> **2** Individual answers. For example: 5 cm, 5 cm, 5 cm; 2 cm, 3 cm, 10 cm; 1 cm, 7 cm, 7 cm

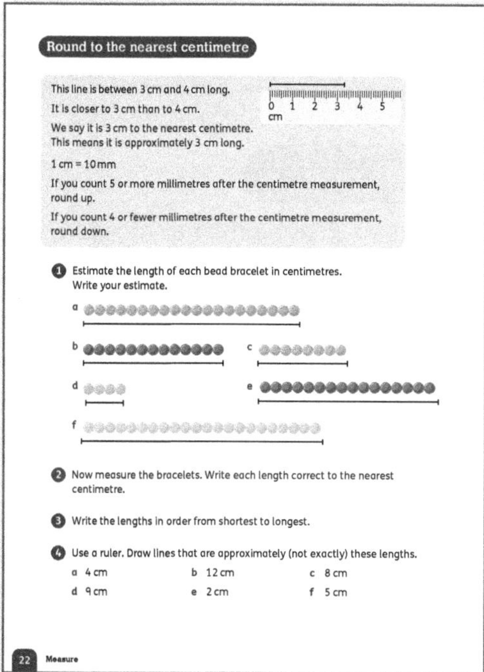

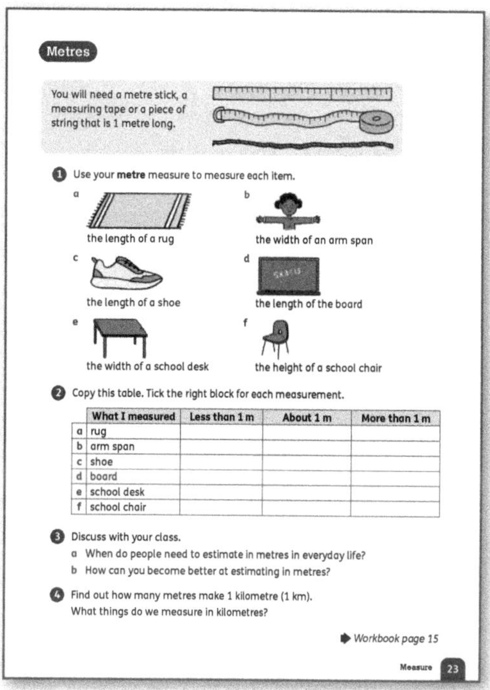

Answers for Pupil Book 3 page 22

Materials

Rulers; number lines; ten frames (see page 21) and counters

Warm-up

- Ask the children to count the number of millimetres in a centimetre.
- Practise rounding single-digit and 2-digit numbers to the nearest 10.

Focus

- Measure different lines and let the children decide which measurement in whole centimetres is the closest to the length of the line. Demonstrate how to *round* to the nearest centimetre, rounding up if the measurement is 5 or more millimetres more than the whole centimetre measurement.
- Let the children notice the similarities between rounding numbers to the nearest 10, and rounding lengths to the nearest centimetre.
- Once you have demonstrated with a few examples, let them work through questions 1–4 on **Pupil Book 3 page 22** independently.

Challenge

Ask the children to draw three lines of different lengths that all round to 10 cm to the nearest centimetre.

Support

Use visual aids such as number lines and ten frames to reinforce rounding to the nearest 10. For example, show 32 on a number line and identify the tens either side: 30 and 40. The children can see that 32 is closer to 30 than to 40.

Metres

Materials

Rulers; metre rules; measuring tapes or pieces of string 1 m long

Warm-up

Put the children into groups and challenge each group to find two things in the classroom that are shorter than their ruler, and two things that are longer than their ruler.

Focus

- Hand out the measuring equipment. Check that the children know how to line up their metre rule against an object to find out if the object is less than, more than or about 1 metre in length.
- Let the children work in pairs or small groups to complete the practical measuring activity in question 1 on **Pupil Book 3 page 23**, recording whether each item is longer, shorter or about the same as 1 metre.
- You may need to take the children through the steps of drawing the table in question 2 in their exercise books. They will need to use rulers to draw the columns and rows, which is also good practice in using the measurements on their rulers.
- Complete question 3 orally with the class.
- Question 4 introduces the idea of a kilometre. The children find out how many metres in 1 kilometre (1000) and suggest things we might measure in kilometres.

Follow-up

Workbook 3 page 15 provides additional practice in estimating and measuring length.

End-of-unit check

- Ask the children to use a ruler to measure some short items to the nearest centimetre. For example, they could measure a shoelace, a lollipop stick and strips of paper.
- Draw some composite paths like those on **Pupil Book 3 page 21** and ask the children to measure them using a ruler. Ask them to find different ways to draw a path with a given total length.

Ask questions such as:

- *Name some things we measure in metres rather than centimetres.* (Possible answers: the length and height of the classroom wall, the length of a fence, the height of a skyscraper.)
- *Which unit (metres or centimetres) would you choose to measure the length of an aircraft wing?* (metres) *the length of the wire from a kettle?* (centimetres) *the height of a light switch from the floor?* (centimetres) *the height of a tree or a lamp post?* (metres)

UNIT 4
Patterns and sequences

Learning objectives
- Use a shape or object to represent an unknown quantity in addition and subtraction.
- Recognise and extend linear sequences and describe the term-to-term rule.
- Extend spatial patterns formed from adding and subtracting a constant.
- Solve problems, including missing number problems, using number facts, place value, and more complex addition and subtraction.

Key words
count on   count back   pattern   sequence   term   term-to-term rule

Unit introduction

Materials
Counters or Numicon number frames (see pages 21–22)

Teaching guidance
- Use counters or Numicon number frames to set out groups of three, arranged in triangles like this:

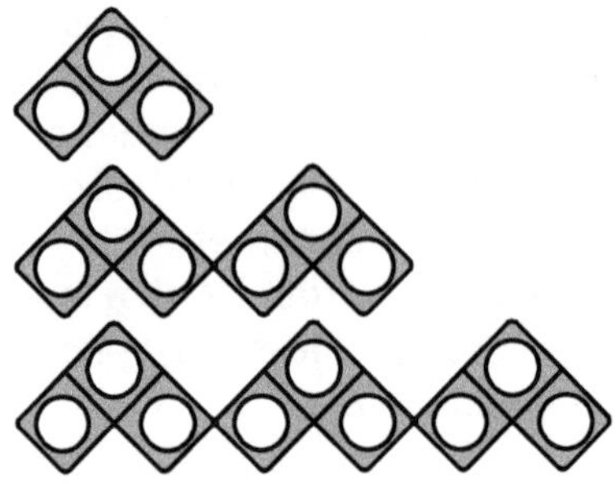

- Ask the children to describe the *pattern*. Ask:
  - *What shapes do you see?*
  - *Does the pattern grow each time?*
  - *What will the next line look like?*
  - *What number would you use to describe each line?*
- The children may notice that each triangle has three circles. So, they may count 1, 2, 3 (triangles) or 3, 6, 9 (circles). Let a child come up to show you the next line in the sequence.
- Let the children use counters to make their own patterns for given number sequences. For example, ask them: *How could you use counters to make this pattern: 4, 8, 12, 16?*

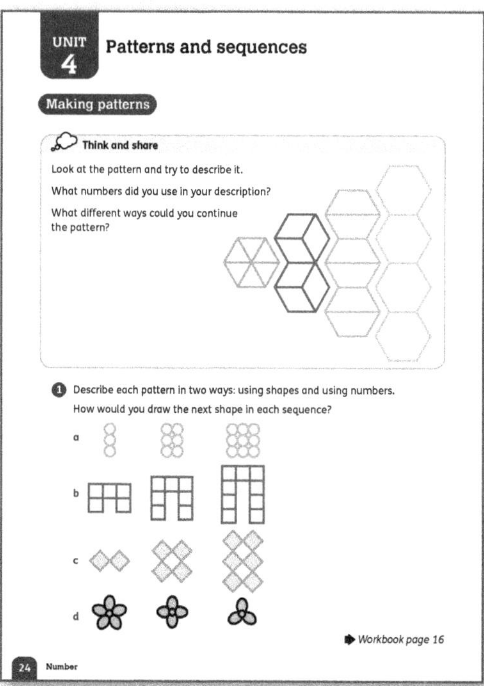

Materials

Counters; square tiles

Warm-up

- <u>Think and share:</u> Ask the children to think about the pattern shown at the top of **Pupil Book 3 page 24**:
  - *What different patterns do you notice?*
  - *How does the pattern grow?*
  - *What changes and what stays the same?*
- The children can work in pairs or groups to discuss these questions.
- There are different ways of seeing and describing the pattern. Let the children explore different ways in which the pattern might continue.
- You might want to suggest one way of continuing the pattern, and then ask the children if they can think of others (see the Answers section for this page, right, for ideas). All suggestions are valid here – the question is open-ended and designed to encourage many different possibilities.

Focus

- In the same pairs or groups, the children describe each pattern in question 1. If necessary, remind them that when we describe something we say what it is, what it looks like, how it works. You may need to ask guiding questions such as:
  - *What is the first shape? How does the second shape change?*
  - *What changes? What stays the same?*
  - *How does the third shape change?*
  - *How would you make the next shape?*

Follow-up

The children work through the patterns on **Workbook 3 page 16**. They also have an opportunity to create their own patterns. Remind them that they can create patterns by adding or subtracting.

Answers for Pupil Book 3 page 24

<u>Think and share:</u> Possible answers:

- one hexagon, two hexagons, three hexagons, four hexagons
- add one hexagon each time
- sixths, thirds, halves, wholes

Children can use numbers to describe the number of hexagons in each column and/or fractions to describe the parts each hexagon is divided into.

Two possible continuations for the pattern:

1 Possible answers:

  a Columns of 3 circles; one more column is added each time
  3, 6, 9 then 12

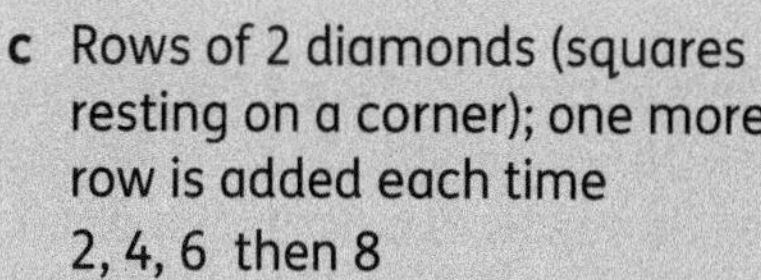

  b 'Bridges' made of squares with three along the top and two 'legs'; two more squares are added to the 'legs' each time
  5, 7, 9 then 11

  c Rows of 2 diamonds (squares resting on a corner); one more row is added each time
  2, 4, 6 then 8

  d Flowers one less petal each time 5, 4, 3 then 2 (a flower with 2 petals)

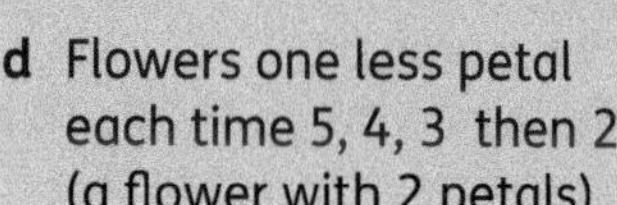

Answers for Workbook 3 page 16

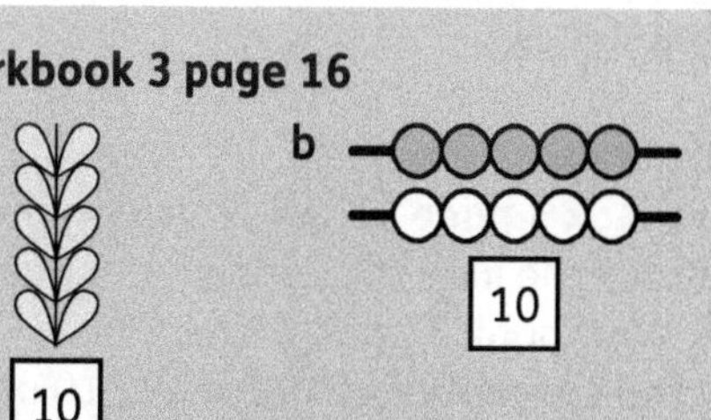

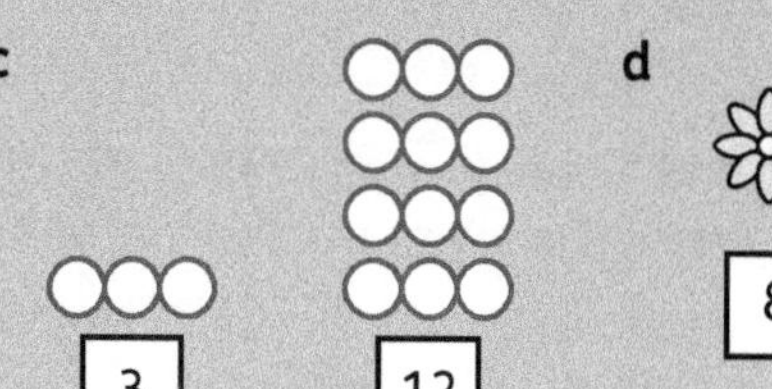

2 Individual answers

Number sequences

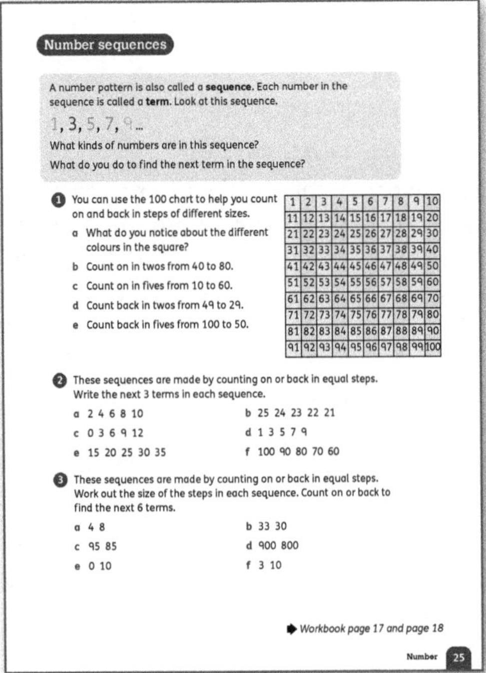

Materials

100 chart

Warm-up

Use 'Number sequences', 'Simple function machines' or
'Function machines' (page 31) to introduce the terms
sequence, *rule* and *term*.

Focus

- Discuss the example at the top of **Pupil Book 3 page 25**.
- Make sure the children understand that a *sequence* is
  a number pattern that goes in a particular order and
  follows a rule. They need to understand that each
  number in the sequence is called a *term*, and that
  each term is made by applying a rule to the term
  before. We can work out the rule by noticing how the
  terms change from one to the next.
- All the sequences in this lesson are made by *counting
  on* and *counting back* in steps of the same size.
- Work through the questions with the class. Let the
  children finish some of the sequences independently
  if possible.

Follow-up

Workbook 3 page 17 uses function machines to
reinforce the idea of changing a number according to a
rule. **Workbook 3 page 18** reinforces counting patterns
and the concept of identifying 1, 10 or 100 more or less
than a given number.

Challenge

'Missing numbers' and 'Count in steps' (page 27) offer
some fun challenges working with number patterns and
sequences.

You can also give the children the following challenge:
*Create a pattern that has only even terms. But your rule
must not be 'add 2' or 'subtract 2'. How many different
ways can you do it?*

Answers for Pupil Book 3 page 25

1 a Possible answers: The yellow numbers are odd
      and the blue numbers are even. This creates
      a pattern of stripes. In both sets of numbers,
      the numbers go up by 2 each time.
   b 40, 42, 44, 46, 48, 50, 52, 54, 56, 58, 60, 62,
      64, 66, 68, 70, 72, 74, 76, 78, 80
   c 10, 15, 20, 25, 30, 35, 40, 45, 50, 55, 60
   d 49, 47, 45, 43, 41, 39, 37, 35, 33, 31, 29
   e 100, 95, 90, 85, 80, 75, 70, 65, 60, 55, 50

2 a 12, 14, 16   b 20, 19, 18
   c 15, 18, 21   d 11, 13, 15
   e 40, 45, 50   f 50, 40, 30

3 a Step size: 4   Next 6 terms: 12, 16, 20, 24, 28, 32
   b Step size: 3   Next 6 terms: 27, 24, 21, 18, 15, 12
   c Step size: 10
      Next 6 terms: 75, 65, 55, 45, 35, 25
   d Step size: 100
      Next 6 terms: 700, 600, 500, 400, 300, 200
   e Step size: 10
      Next 6 terms: 20, 30, 40, 50, 60, 70
   f Step size: 7   Next 6 terms: 17, 24, 31, 38, 45, 52

Answers for Workbook 3 page 17

1 a 99   b 77, 95, 32
   c 14 → 13, 29 → 28, 77 → 76
   d 41 → 142, 207 → 307, 485 → 585
   e 750, 421, 600   f 731, 595, 390
   g 388 → 488, 460 → 560, 25 → 125
   h 199 → 198, 428 → 427, 539 → 538

Answers for Workbook 3 page 18

1 a 35   b 82   c 100
   d 26   e 87   f 39
2 a 135   b 270   c 367, 368, 369
3 a 336   b 451   c 640   d 790
4 a 145   b 260   c 349   d 499
5 a 990   b 844   c 699   d 672
6 a 281   b 335   c 690   d 663
7 a fifty-one b one hundred and forty
   c four hundred and ten   d seven hundred
8 a 600   b 700   c 440
   d 550   e 1000   f 135

Term-to-term rules

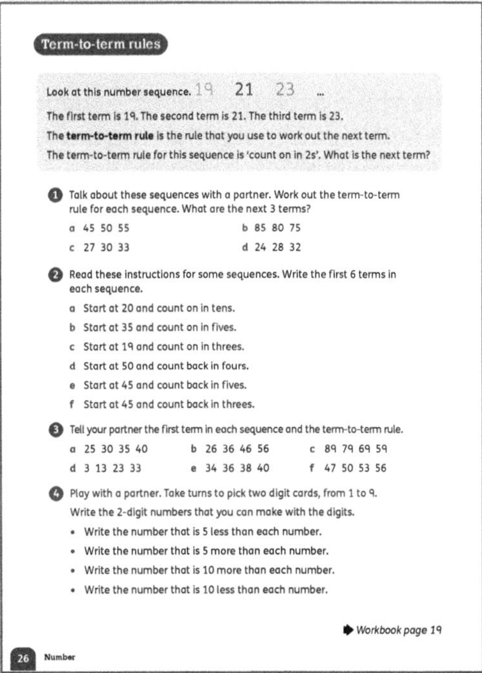

Materials
Two sets of 1–9 digit cards for each pair of children

Warm-up
- Use the example at the top of **Pupil Book 3 page 26** to introduce the concept of a *term-to-term rule*. You can explain that this is the special name we use to talk about the rule that takes us from one term to the next.
- It is important to point out to the children that once we have worked out a rule, we can test it to make sure it works on all the numbers in the same way. For example, show the children these patterns:
  ◦ 1, 2, 1, 2, 1, 2, …
  ◦ 1, 2, 3, 4, 5, …
- Ask: *Do they both follow the same rule? If I look at the first two terms in each sequence, I think the rule is 'add 1'. Is that the rule for both patterns? Why not?*
- The number sequence in the example in the **Pupil Book** follows a simple +2 rule. You might want to contrast this with another sequence that starts with the same terms but then changes:
  19, 21, 24, 28, … (add 2, then add 3, then add 4 …).

Focus
- Let the children work in pairs to identify the term-to-term rule and the next three terms in each sequence in question 1.
- Question 2 gives the first term and the rule. The children write the first six numbers in the sequence.
- In question 3, the children identify the first term and the rule.
- The children play a game in pairs in question 4, using two sets of 1–9 digit cards and writing the numbers they generate by applying the given rules.

Follow-up
Number sequences are closely related to counting and skip counting. The children will notice particular patterns as they repeatedly add or subtract ones, fives, tens and so on. The skill of following a given 'function' instruction, and noticing the patterns, is consolidated further on **Workbook 3 page 19**.

Support
You can draw some function machines for the children to complete. They can work out the missing numbers in or out, or the rules. Use the simple number machines on **Workbook 3 page 26** (Revise number facts) as a template to make more examples.

Answers for Pupil Book 3 page 26
1 a Count on in 5s   Next 3 terms: 60, 65, 70
   b Count back in 5s   Next 3 terms: 70, 65, 60
   c Count on in 3s   Next 3 terms: 36, 39, 42
   d Count on in 4s   Next 3 terms: 36, 40, 44
2 a 20, 30, 40, 50, 60, 70   b 35, 40, 45, 50, 55, 60
   c 19, 22, 25, 28, 31, 34   d 50, 46, 42, 38, 34, 30
   e 45, 40, 35, 30, 25, 20   f 45, 42, 39, 36, 33, 30
3 a Start at 25 and count on in 5s.
   b Start at 26 and count on in 10s.
   c Start at 89 and count back in 10s.
   d Start at 3 and count on in 10s.
   e Start at 34 and count on in 2s.
   f Start at 47 and count on in 3s.
4 Individual answers
   If the two cards show the same digit, then there is only one possible 2-digit number.

Answers for Workbook 3 page 19
1 a 424 (Provided as an example)   b 217
   c 551   d 683   e 936   f 705
2 a on in 7s (Provided as an example) b on in 1s
   c back in 1s   d on in 10s
   e back in 10s   f on in 1s
   g on in 100s   h back in 10s
3 a 665   b 409   c 1009
4 a 865   b 590   c 739
5 a 300   b 239   c 750
6 a 400   b 450   c 390
7 a back   b back   c on
   d back   e on   f on

Using shapes and objects

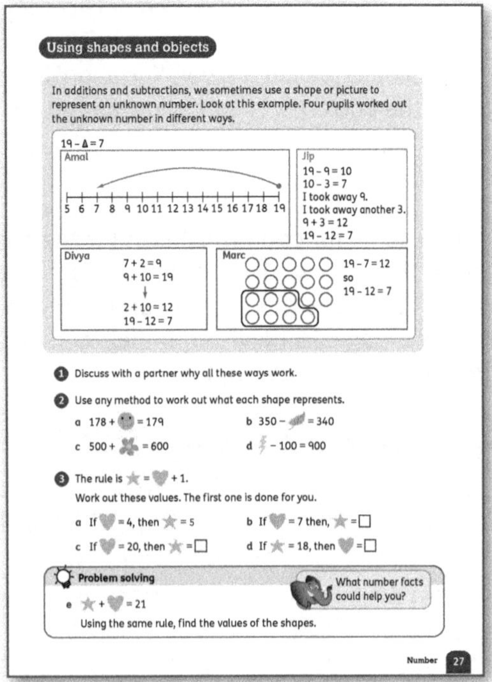

Materials

Small stones or counters

Warm-up

- Draw this number sentence on the board:

$$\heartsuit + \triangle = 10$$

- Explain that each shape represents a number. Say: *We know that adding the numbers together makes 10. What number could the heart be worth?*
- Let them make suggestions, for example 3. Ask: *If the heart is worth 3, what is the triangle worth?* (7)
- Explain that in a number sentence like this, there are some different possible values for each shape. Let the class work through them, noticing that they are using number bonds to 10 to work out the possible values.
- You can try this with bonds to different numbers and using different shapes, for example:

$$\text{leaf} + \text{flower} = 12$$

- You might also want to try this with higher numbers. When working with 3-digit numbers, it is not necessary to work through every possible combination of values.

Focus

- Look at the example at the top of **Pupil Book 3 page 27** with the class:

$$19 - \triangle = 7.$$

- The children have already seen some shapes used to represent numbers. Discuss what the triangle in the example represents.
- Let the children work in pairs to answer questions 1 and 2. Give them small stones or counters to help them.
- Point out that the shapes in question 3 have a different value each time.
- <u>Problem solving</u>: In question 3e, the children find all the possible pairs of values that add to 21.

Answers for Pupil Book 3 page 27

1 Possible answers:

Amal counted back to 7 on the number line to find out the number to take away from 19 to make 7.

Jip used 'chunking'. He worked back from 19 to 10 and from 10 to 7 using subtraction. He added together the 'chunks' he took away to find the answer.

Divya also used 'chunking'. She worked up from 7 to 9 and and then from 9 to 19 using addition. She added up the 'chunks' she added to find the answer.

Marc used counters. He took away 7 counters from 19 counters to find the answer.

2 **a** 1   **b** 10   **c** 100   **d** 1000

3 **a** star = 5 (Provided as an example) **b** star = 8
   c star = 21   **d** heart = 17
   e star = 1, heart = 20; star = 2, heart = 19; and so on to star = 20, heart = 1

End-of-unit check

Check the children's understanding of this unit by asking them to:

- write their own rules for number patterns and give them to a partner to write the number patterns (Individual answers. For example: Start at 1 and count on in fives. 1, 6, 11, 16, …)
- create a shape pattern that is based on a given number sequence, 3, 6, 9, …. (Individual answers. For example: a pattern with 3 triangles, 6 triangles, 9 triangles, …).

Give the children number or shape patterns and ask them to give the next three terms and to tell you the term-to-term rule. For example, 6, 10, 14, 18, … (22, 26, 30; term-to-term rule: 'add 4')

Ask the children to find the value of the star in equations such as $57 + \star = 60$ (3) and $140 - \star = 120$ (20).

Learning objectives
- Recognise an angle as a description of a turn.
- Identify right angles.
- Recognise that a straight line is equivalent to two right angles or a half turn.
- Identify horizontal and vertical lines and pairs of perpendicular and parallel lines.
- Recognise that two right angles make a half turn, three make three-quarters of a turn and four make a complete turn.
- Identify whether angles are greater than or less than a right angle.

Key words
angle   right angle   half turn   full turn
quarter turn   three-quarter turn   square corner
90-degree angle   straight line   parallel lines
perpendicular lines   horizontal lines   vertical lines

Unit introduction

Materials
Moveable angles (made by joining two strips of card with a paper fastener or split pin); pictures of objects that have angles, such as sketches of a variety of chairs with the back and seat at different angles or ladders open to different angles – you can sketch these on the board

Teaching guidance
- Without saying anything, hold up a moveable angle to show the class. Show it in a closed position. Slowly start opening it to form different angles less than a right angle. Pause, holding up an angle and ask the class to guess: *What do you think I'm showing you?*
- Let the children guess or make suggestions. They might see it as the hands of a clock, or they might already remember the word *angle*. They might say you are opening up the strips or hands.
- Ask: *What will happen if I continue? What shape will the two lines make?*
- Demonstrate how the two strips can form a right angle. The children may suggest that it is a *square corner,* a *90-degree angle*, an 'L-shape' or they may recall the term *right angle* or *quarter turn*.
- Continue opening the strips and asking: *What will happen if I keep opening the lines?* Open the strips further to form a *straight line* (a *half turn*), then a *three-quarter turn,* then to the closed position again *full turn*.
- Let the children discuss what we call two lines that meet at a point (an angle). Explain that an angle is a measure of turn. You may want to sketch some angles on the board and have the children draw larger or smaller angles.

- Remind the children of the term *right angle* and let them demonstrate some right angles, both by using their moveable angles and by showing you objects with right angles.

Challenge
Show the children a city street map and have them identify angles that are greater than, less than or equal to a right angle on the map.

Support
Some children may need help to understand what we mean by an angle. You can use different practical ways to help them develop understanding of this concept:
- using drawings of lines
- opening and closing a pair of scissors
- opening and closing a pair of joined moveable strips
- opening and closing a door
- using moveable clock hands.

It is important that the children grasp that the angle is created about the point where the two lines join. It is the measure of the opening or turn from one line to the other.

You can also use cut-out shapes and show the children the corners. Another way to think of an angle is as a corner. The angle tells us how much to turn around the corner.

Interesting mistakes
You may need to explain that a 'larger angle' refers to the size of the turn, not the size of the sketch. So, the following two angles are the same size:

Show the children pictures of real-life angles (such as in chairs and ladders) and let them discuss which angles are smallest (closest to being closed) and which are largest (most open).

Have the children demonstrate different angles using their arms or using objects on the desk.

Angles

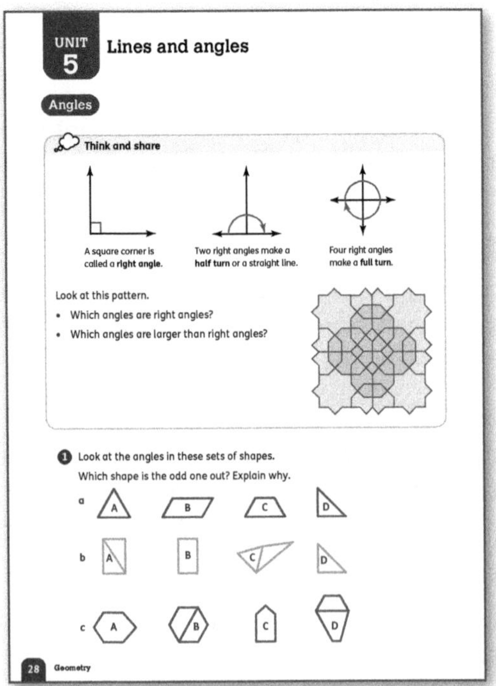

Materials

Some objects that have right angles, such as a book, a piece of paper, a box

Warm-up

- Have the children identify right angles in the classroom and on the objects you have provided.
- Explain that a right angle can tell us how two lines meet at a point. It can also tell us how far one line has turned or opened away from another (a degree of turn).

Focus

- <u>Think and share:</u> Discuss the pictures of angles at the top of **Pupil Book 3 page 28**.
- Then let the children work in groups or pairs to talk about what angles and shapes they can see in the pattern. Let them identify right angles, angles that are larger than right angles and different shapes. They may see stars, pentagons, hexagons and squares.
- Let the children continue working in their pairs or groups to discuss the 'odd one out' in each set of shapes in question 1. This is intended to be a challenging question, and the children may disagree on which shape doesn't belong. Accept any answers based on sound reasoning. The idea is that they get used to looking at the sides and angles of each shape, and noticing what does and does not match.

Answers for Pupil Book 3 page 28

<u>Think and share:</u> Example answers (right angles marked with the square symbol; angles larger than right angles marked with curves)

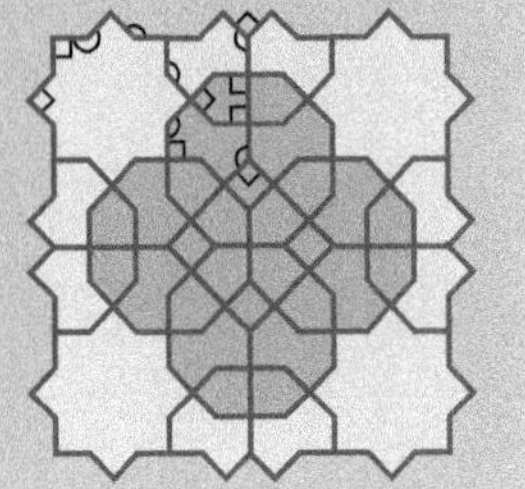

1 Possible answers:

   a D because it is the only shape with a right angle.

   A because it is the only shape with all sides of equal length.

   b C because it is the only shape without any right angles.

   B because it is the only shape with no triangles.

   c C because it is the only shape with right angles.

   C because it is the only pentagon (the others are hexagons).

Right angles

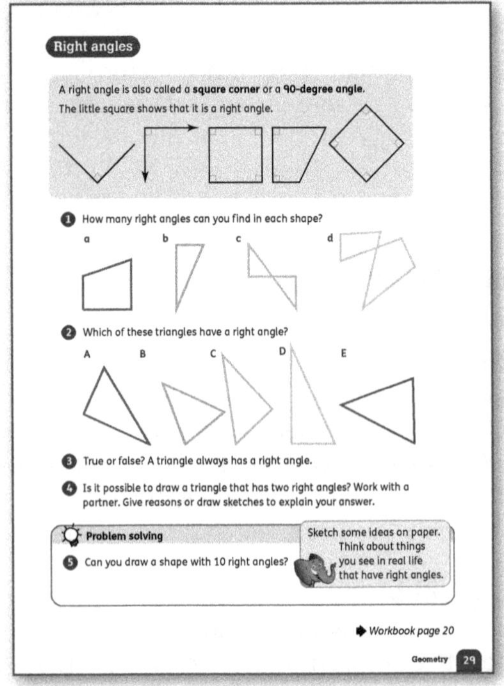

Materials

Rulers and set squares

Warm-up

- Discuss the right angles shown at the top of **Pupil Book 3 page 29**.
- Explain that the position of the angle doesn't change its size. Demonstrate this by turning a piece of paper or a book with a right-angled corner.

Focus

- Demonstrate that we mark a right angle using a small square corner.
- Let the children work in pairs or independently through questions 1–4.
- <u>Problem solving:</u> The children try to draw a shape with 10 right angles, using rulers and set squares. Point out that the shape can include other angles as well.

Follow-up
Workbook 3 page 20 consolidates the work on right angles.

Interesting mistakes
When the children identify angles incorrectly, let them explain why they thought the angle was that size. Sometimes it is more difficult to identify a right angle when it is in an unusual position. Invite the children to suggest ways of checking the size of an angle, for example using the corner of a piece of paper.

Answers for Pupil Book 3 page 29
1 a 2   b 1   c 2   d 2
2 A, C, D
3 False
4 No, it is not possible. If you draw a line for the base of the triangle and draw a right angle at each end, the lines will never meet.
5 Individual answers. For example:

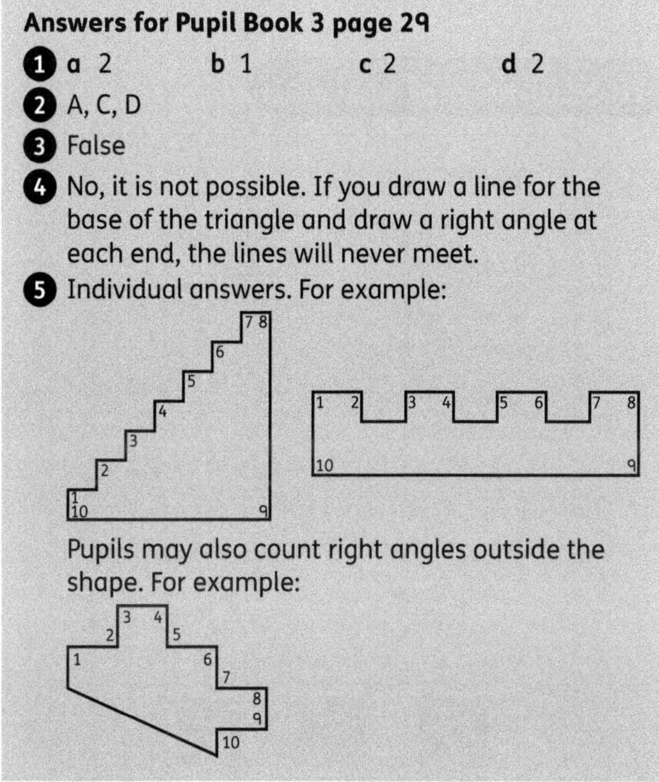

Pupils may also count right angles outside the shape. For example:

Answers for Workbook 3 page 20
1 a

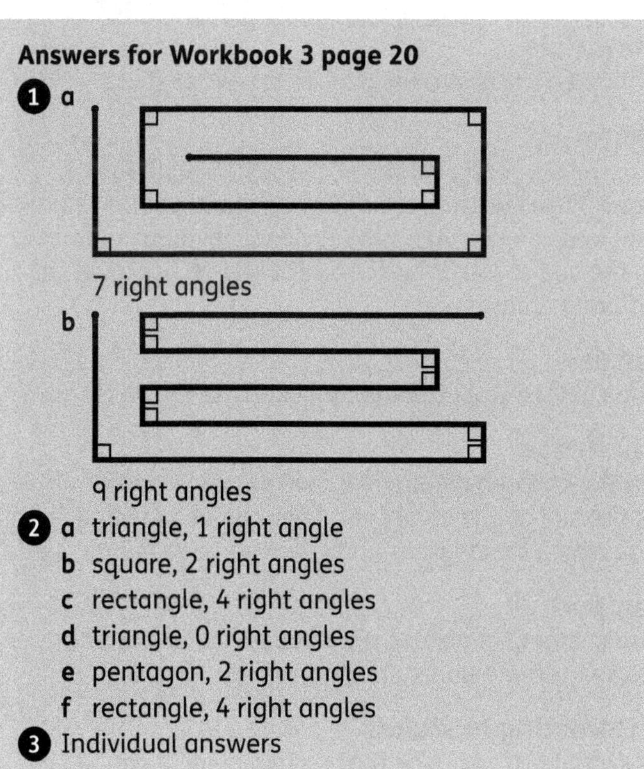

7 right angles

b

9 right angles

2 a triangle, 1 right angle
  b square, 2 right angles
  c rectangle, 4 right angles
  d triangle, 0 right angles
  e pentagon, 2 right angles
  f rectangle, 4 right angles
3 Individual answers

Parallel and perpendicular lines

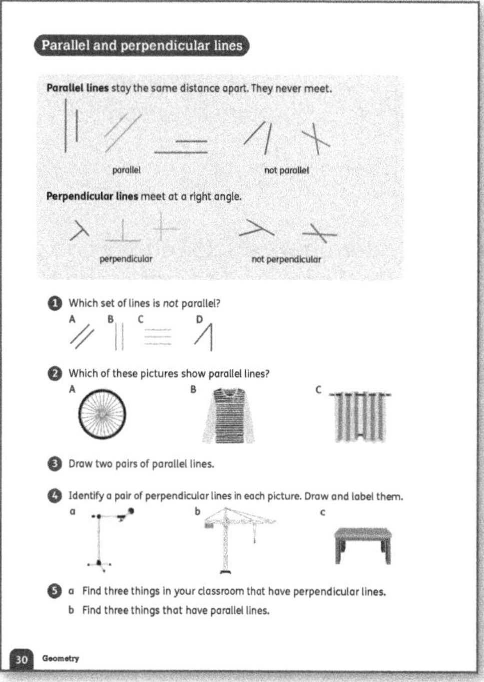

Materials
Rectangular and square objects such as books, rulers, blocks, notepads; a large ruler; pencils, straws or sticks

Warm-up
- Hold up a rectangular object such as a book or ruler. Point to the two parallel long sides and ask the class: *What do you notice about these two lines?* They may notice that they are the same length.
- Draw a rectangle on the board, then draw again over the two parallel long sides to emphasise them. Say: *Imagine we continue these lines up and down. Will they ever meet? Why not? What can you say about these lines?*

- Let the children talk about the positions of the lines in their own words.

Focus
- Introduce the term *parallel lines* and write it on the board. Explain that parallel lines never meet. They stay the same distance apart.
- Use a ruler to draw some different pairs of lines on the board (some pairs parallel, others not). Let the children identify which are parallel and which are not.

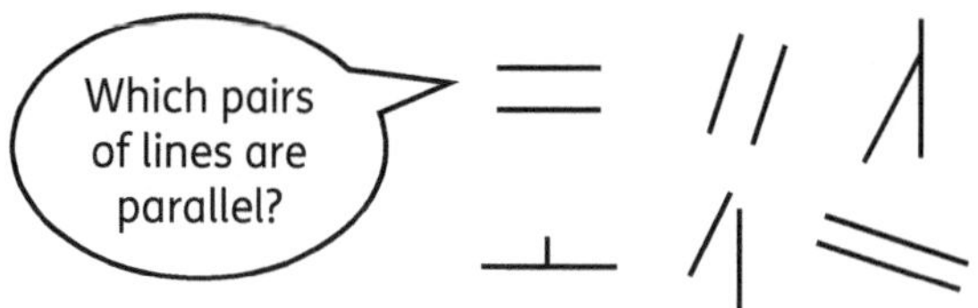

- Point out the pair of short parallel lines on the rectangle. Ask: *Are these lines parallel too? How do you know?*

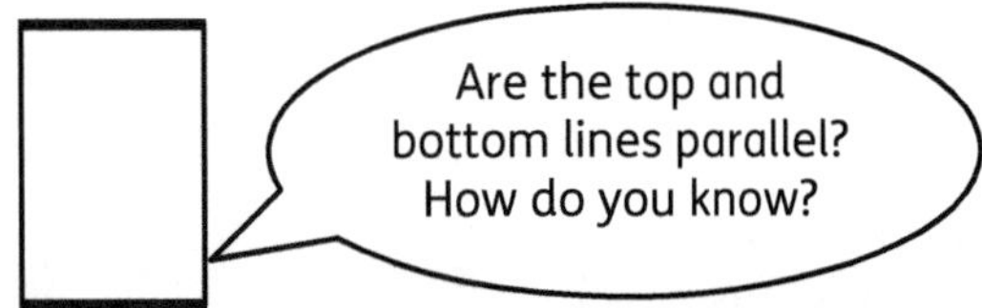

- Once the children have discussed the two opposite pairs of parallel lines, point to one of the long sides and the short side at the bottom of the rectangle. Ask: *What about these two lines? Are they parallel? Why not?* Let the children explain that these lines are not parallel because they meet.

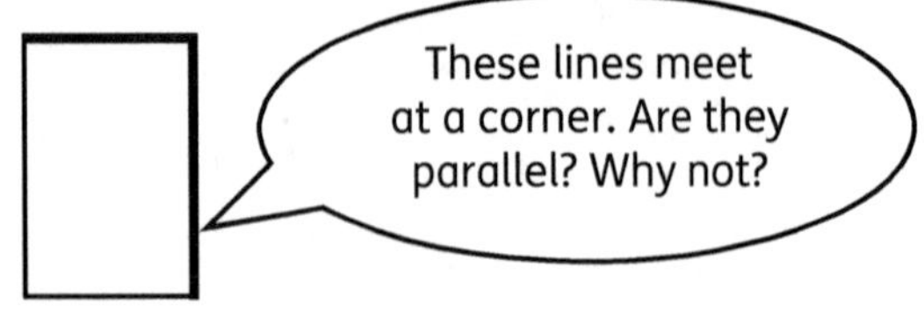

- Ask: *Do you notice something special about the angle at which these lines meet?* If necessary, point out that these are square corners (right angles). Introduce the term *perpendicular lines* and write it on the board. Explain that perpendicular lines meet at a right angle.
- Again, draw some pairs of lines – some perpendicular, some not – and have the children identify which are perpendicular. Let them explain why. You might also let them draw some examples on the board.

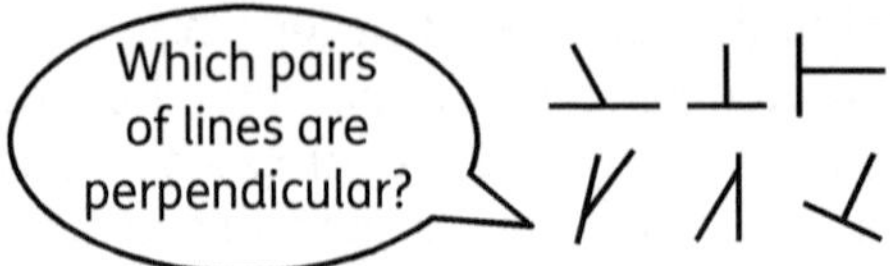

- The children can work independently through questions 1–5 on **Pupil Book 3 page 30**.

Challenge
The children can draw their own patterns using parallel lines.

Support
Use objects such as pencils, straws or sticks to create pairs of perpendicular lines and pairs of parallel lines. Ask the children to identify whether each pair is perpendicular or parallel and to use another pair of pencils to make the same kind. Gradually progress to showing pairs of parallel and perpendicular lines in different positions.

Answers for Pupil Book 3 page 30
1. D
2. B and C
3. Individual answers

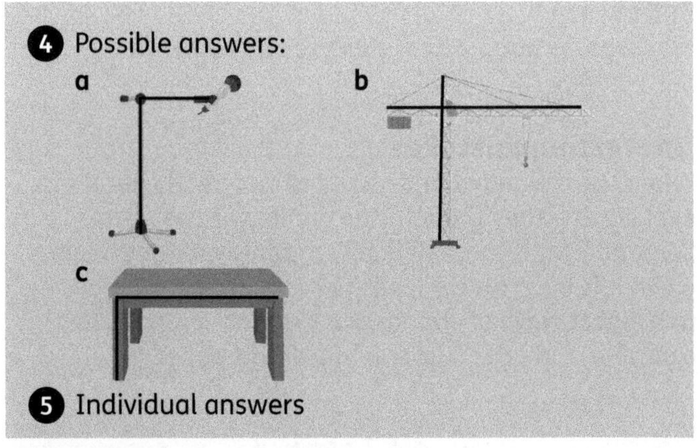

4. Possible answers:
a
b
c

5. Individual answers

Horizontal and vertical

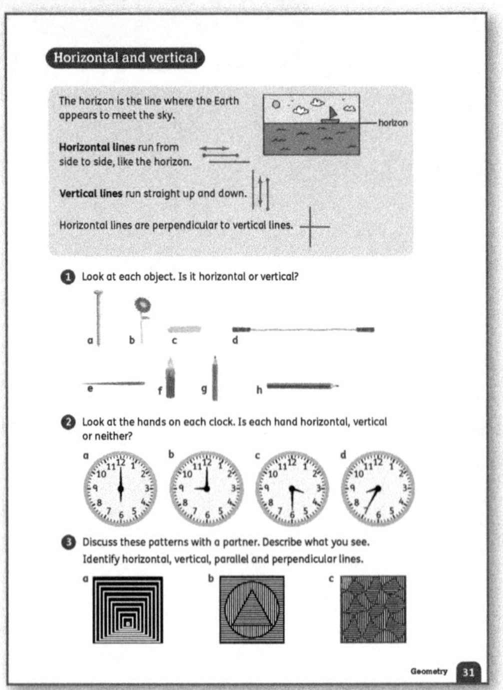

Materials
Straight objects such as pencils, straws or sticks

Warm-up
Use the text and pictures at the top of **Pupil Book 3 page 31** to illustrate the difference between *horizontal ines* and *vertical lines*. Let individual children come up to the board and draw horizontal and vertical lines of different lengths.

Focus
The children work through questions 1–4 in pairs.

Challenge
Let the children create their own patterns using horizontal and vertical lines. They can use the images in question 3 for ideas.

Support
Use a variety of objects such as pencils, straws and sticks to create horizontal and vertical lines.

Interesting mistakes
The children may think that certain objects are intrinsically horizontal or vertical. It is important to show that these words describe position, and that position

can change. For example, show them that a ruler can be horizontal when positioned one way and vertical when positioned differently.

Answers for Pupil Book 3 page 31

1 **a** vertical   **b** vertical   **c** horizontal
   d horizontal   **e** horizontal   **f** vertical
   g vertical   **h** horizontal

2 **a** both vertical   **b** one vertical, one horizontal
   c one vertical   **d** neither

3 Possible answers:
   a Squares with horizontal and vertical sides, each square inside a larger square
   b A triangle filled with horizontal lines inside a circle filled with vertical lines inside a square filled with vertical lines
   c A pattern of shapes, each with four curved sides. The shapes are alternately filled with horizontal and vertical lines.

Turns

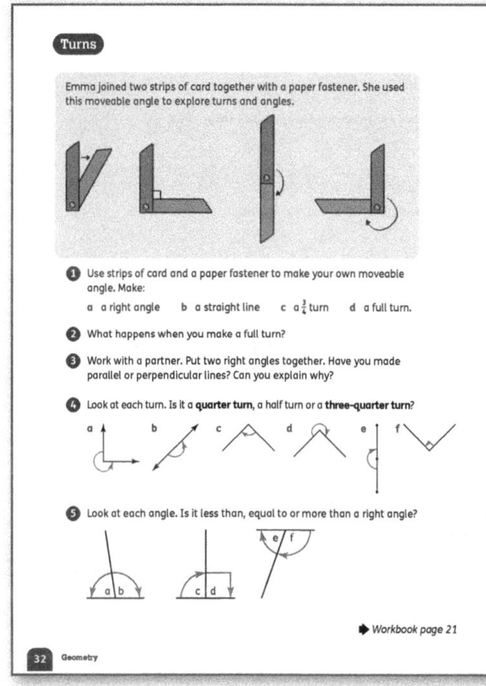

Materials

Strips of card and paper fasteners or split pins (for the children to make their own moveable angles)

Warm-up

Supervise the class to make their own moveable angles. Let the children demonstrate different turns. Relate the different turns to right angles:

- one right angle = a quarter turn
- two right angles = a half turn
- three right angles = a three-quarter turn
- four right angles = a full turn.

Focus

- The children work through questions 1–5 on **Pupil Book 3 page 32**.

- For questions 4 and 5, explain that the arc shows the angle they are interested in. For example, in 4a the turn they need to describe is the three-quarter turn marked with the arc.

Follow-up

Workbook 3 page 21 consolidates the work on lines and angles from this unit. If you wish, you can use this page as an informal assessment activity.

Answers for Pupil Book 3 page 32

1 Practical work

2 The two strips close up again./They lie one on the other.

3 Perpendicular. Possible explanation: Two of the lines make a straight line with the other line perpendicular to it./Perpendicular lines join at a right angle.

4 **a** three-quarter turn   **b** half turn
   c quarter turn   **d** three-quarter turn
   e half turn   **f** quarter turn

5 **a** less than   **b** more than
   c equal to   **d** equal to
   e less than   **f** more than

Answers for Workbook 3 page 21

1 Individual answers

End-of-unit check

Recap with the children what they have learnt in this unit.

- Let them use their moveable angles to show you:
  - a *right* angle
  - an angle greater than a right angle
  - an angle smaller than a right angle.
  An example of each type of angle is shown here.

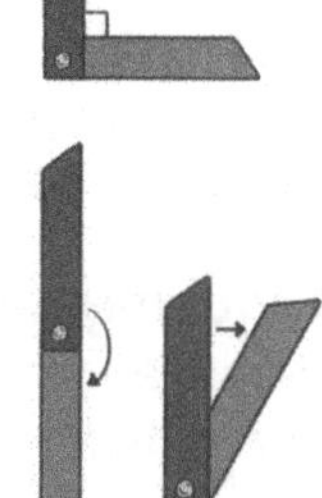
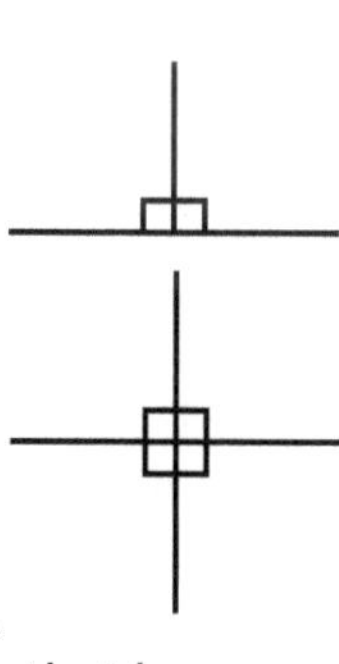

Then ask:

- *How many right angles make a straight line?* (2)
- *How can I show this using a drawing or moveable angles?*
- *How many right angles make a full turn?* (4)
- *How can I show this?*

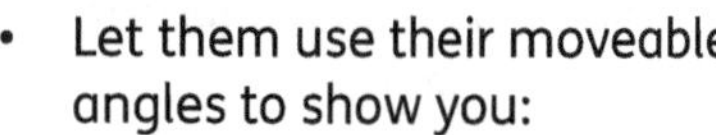

- Show the children a variety of shapes. *Show me the right angles in these shapes.*
- *Show me some objects in the classroom that have parallel and perpendicular lines.* (Possible answers: the edges of square and rectangular objects, such as the whiteboard, exercise books, desks, tables, windows and doors.)
- *Show me some vertical and horizontal lines in the classroom.*
- Show the children some drawn angles and ask if each one is less than, equal to, or more than a right angle.

Learning objectives
- Identify, describe, classify, name and sketch 2D shapes by their properties.
- Recognise angles as a property of shape.
- Differentiate between regular and irregular polygons.
- Draw polygons by joining marked points.
- Identify parallel and perpendicular sides of polygons.
- Identify right angles in 2D shapes presented in different orientations.
- Identify horizontal and vertical lines of symmetry on 2D shapes and patterns.
- Sketch the reflection of a 2D shape in a horizontal or vertical mirror line.

Key words
polygon   side   vertex (vertices)   square rectangle   triangle   pentagon   hexagon quadrilateral   octagon   regular polygon irregular polygon   circle   half   quarter degrees   protractor   compasses   set square ruler   symmetry   line of symmetry   symmetrical

Unit introduction

Materials
A set of shape domino cards (you can search online for these); shape flashcards; cut-out 2D shapes

Teaching guidance
Here are some ideas for revising the 2D shapes and their names before you start teaching this unit.

- The children can play a game of shape dominoes. (Some sets of shape dominoes include a circle. As this unit is about polygons, you may wish to adapt your set to include polygons only.)
  Arrange the children in groups of four and give each group a set of shape domino cards. The children shuffle the domino cards and deal seven cards to each child. One child places a card on the table. The children then take turns to match either a shape to its name or a name to its shape. (Decide on a rule for who places the first domino on the table.)
- Use flashcards with pictures of the shapes on them. Show a flashcard and ask the children to name the shape. They must say how they decided which shape name to use (for example, 'It has four sides and all the sides are equal, so it is a square.').
- Allow the children to spend some time sorting and classifying a range of 2D shapes. Ask the children to find all the shapes with four sides, all the shapes with more than four sides, and so on.

Support
Display six 2D shapes labelled A to F. Display the correct names of the shapes in a mixed order and ask the children to match them up.

Polygons

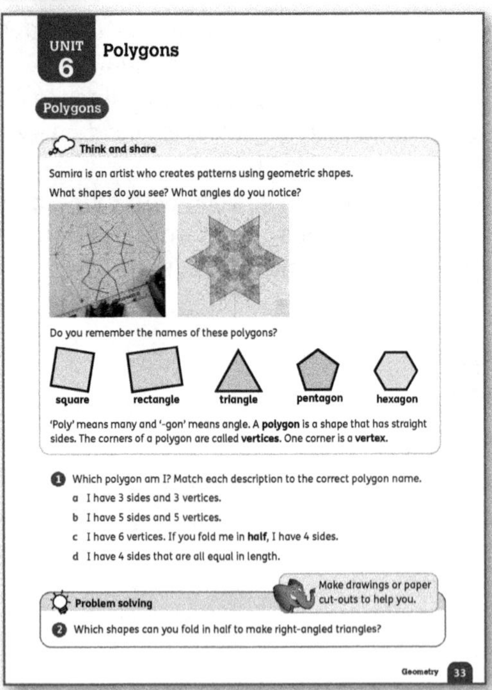

Materials
Cut-out 2D shapes – regular and irregular polygons (triangles, quadrilaterals, pentagons, hexagons, etc.), circles

Warm-up
<u>Think and share:</u> Turn to **Pupil Book 3 page 33** and discuss Samira's artwork with the class. Let the children talk about the shapes and angles they can see.

Focus
- Discuss the *polygons* shown on the page. Explain that the *sides* of polygons are straight lines and we name polygons according to the number of sides.
- Each corner of a polygon is called a *vertex* (plural: *vertices*).
- Check that the children understand that polygons only have straight sides by asking questions such as: *Is a circle a polygon? Why not?*
- Once you have revised the names of the basic polygons (*square, rectangle, triangle, pentagon, hexagon*), let the children answer question 1.
- <u>Problem solving:</u> In question 2, the children can use cut-out 2D shapes to investigate which shapes will form right-angled triangles when folded in *half*.

Follow-up

The children could carry out some of these activities:

- Challenge the children to draw their own examples of each shape.
- The children can make a scrapbook by sticking pictures of polygons (2D shapes with straight sides) used in designs and patterns cut from magazines and other publications. Ask the children to bring magazines into school prior to this activity.
- The children can cut up one simple polygon to make others. For example, they can cut up a square to make four triangles; a rectangle to make a square and a smaller rectangle.
- The children can cut out and use simple polygons (square, rectangle, triangle) and use them to make compound shapes. Challenge each child to create a picture using only these shapes. Use their pictures as part of a display about polygons.

Interesting mistakes

The children may be confused by squares and rectangles as they are both four-sided shapes, whereas other polygons do not have different names for their regular and irregular versions. In fact, a square is a type of rectangle that has all four of its sides the same length. The word *rectangle* means right angle, because a rectangle is a shape whose angles are all right angles.

Answers for Pupil Book 3 page 33

<u>Think and share:</u> circles, triangles, squares, quadrilaterals, pentagons, stars and a hexagon
Right angles, angles that are less than a right angle, angles that are greater than a right angle

1 a triangle   b pentagon
   c hexagon   d square

2 square, rectangle, triangle (if it is symmetrical)

Regular and irregular polygons

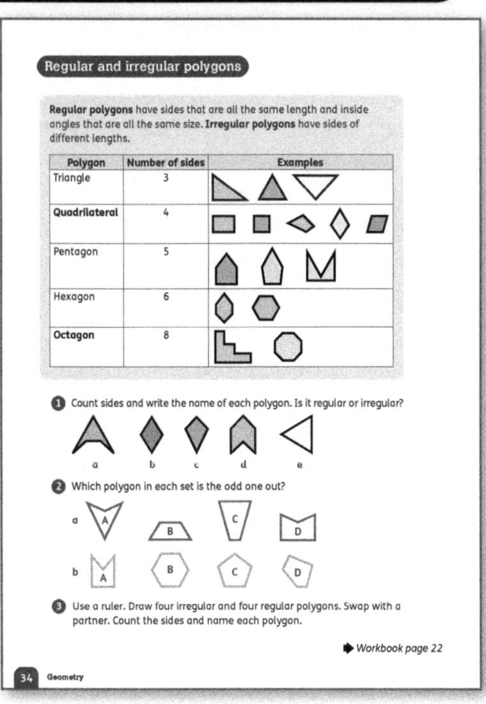

Materials

Cut-out 2D shapes – regular and irregular polygons (triangles, quadrilaterals, pentagons, hexagons, etc.)

Warm-up

- Revise the term *polygon* with the class. Remind the children that the word comes from 'poly' meaning many, and 'gon' meaning side.
- Ask them to suggest and draw examples of some polygons they know (such as squares, rectangles, triangles).

Focus

- Ask who can guess what a *regular polygon* is. Explain that it is a polygon that has all its sides and inside angles equal. Ask: *If a polygon has all sides and angles equal, what do we call a regular four-sided shape or 'quadrilateral'?* (a square)
- Explain that most of the other regular shapes do not have special names like a square – we just call them a regular pentagon, a regular hexagon, and so on.
- Hold up different cut-outs of different shapes and ask the children to say which are regular polygons and which are *irregular polygons*.
- Work through the text and table on **Pupil Book 3 page 34**. Introduce or review the shape names *quadrilateral* and *octagon*. Then let the children work through questions 1–3 independently.

Follow-up

Workbook 3 page 22 provides further practice with 2D shapes, right angles and regular and irregular polygons. Use it for extra practice or assessment.

Challenge

Give the children this challenge: *Try to draw a triangle that has two sides the same length. How can you do this using only a ruler? Try to draw different triangles that have two sides equal in length. What do you notice about the angles?*

Support

The children can work with cut-out shapes to build a given polygon in different ways, for example:

- *Start with a square. What shapes can you add to make it into a triangle?*
- *Start with a hexagon. What shapes can you add to build it into a triangle?*
- *How many different ways can you use triangles to make larger triangles?*
- *How many different shapes can you combine to make a triangle?*

You can also search online for an interactive shapes app.

Here are some examples of triangles:

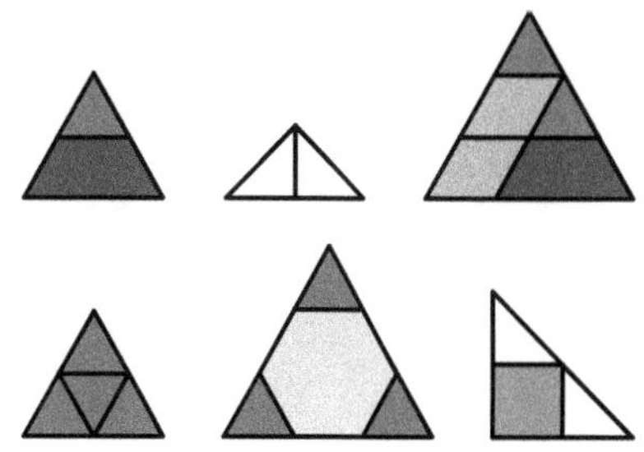

The more the children play with 2D shapes (using their properties practically, naming and combining them), the more comfortable they will become with identifying, classifying and sorting them.

Interesting mistakes

The children might think that irregular polygons with equal sides but different interior angles (such as the rhombus in question 1) are regular. Invite them to check both the sides and the angles.

When building with shapes, the children may sometimes try to add a shape and find that the angles do not work, for example:

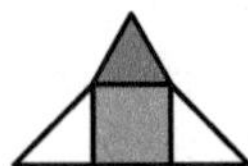

This is a good opportunity to ask: *Why does it not make a triangle? What do you notice? What shape does it make? Can you draw the shape that would make the triangle the way you want it?*

Answers for Pupil Book 3 page 34

1 a 5 sides – irregular pentagon
   b 4 sides – irregular quadrilateral (a rhombus)
   c 4 sides – irregular quadrilateral (a kite)
   d 6 sides – irregular hexagon
   e 3 sides – regular triangle
2 a D is a pentagon; the others are quadrilaterals.
   b B is a hexagon; the others are pentagons.
3 Individual answers

Answers for Workbook 3 page 22

1 A (all four right angles marked)
   E (all four right angles marked)
   J (the right angle marked)

2

	Shape
No right angles	B, C, D, F, G, H, I, K
1 right angle	J
More than 1 right angle	A, E

3 a

Regular polygons	Irregular polygons
A, D, H	C, E, F, G, I, J, K

   b B (the circle) because it is not a polygon – it does not have straight sides or angles.

Angles in shapes

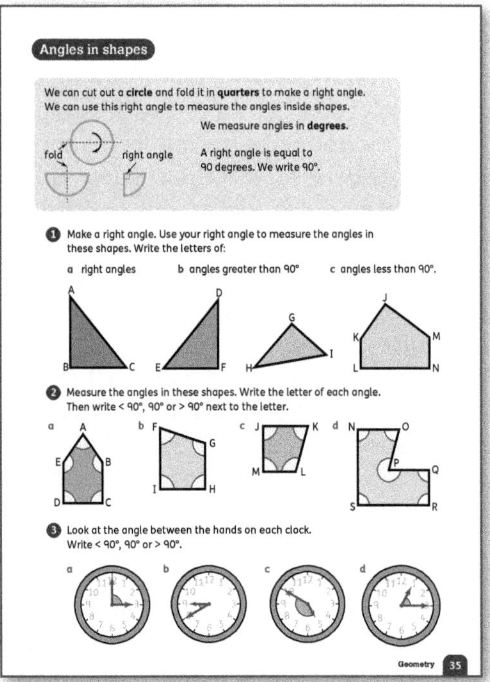

Materials

Cut-out 2D shapes; paper circles; set squares and rulers; large Carroll diagram for display (see Focus)

Warm-up

Give the children a set of cut-out shapes with some right angles (squares, rectangles and right-angled triangles). Show them that we can tear two right-angled corners off a right-angled shape and place the angles along a ruler to show that two right angles make a straight line.

Focus

- Demonstrate how to make a right angle by folding a circle into *quarters*, as shown at the top of **Pupil Book 3 page 35**. Then let each child make their own right angle. Let them spend time using their right angle to test whether angles are right angles or not.
- Revise the terms *greater than* and *smaller than* and the > and < signs as necessary.
- A *set square* is a triangular ruler. Two of the edges are perpendicular. Using a set square to draw a right angle is not as simple as just drawing 'round the corner' of the set square because many set squares have a slight curve on the right-angled corner where they come out of the mould.
- Show the children how to use a set square and a straight edge (ruler) correctly to draw a right angle like this:

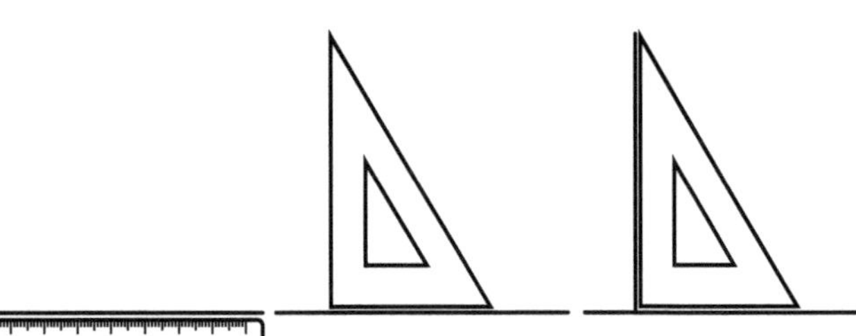

- Let the children try this and make sure they can do it correctly before moving on.
- Prepare a large Carroll diagram like this one for display.

	Has a right angle	Does not have a right angle
Has three sides		
Does not have three sides		

- Show the children a range of 2D shapes. Let them name each shape and then decide where it should go in the Carroll diagram.
- Then let the children work in pairs to complete questions 1–3 on **Pupil Book 3 page 35**.

Challenge
The children can try these challenges:
- *Try making a triangle with the largest possible inside angle you can. What happens to the other angles as you make one of the angles larger? (They become smaller.)*
- *Try making a quadrilateral with the largest possible angle. What different shapes can you make? (The children can make a range of quadrilaterals, including kite, trapezium, parallelogram, rhombus and other quadrilaterals, but they are not expected to know the names at this stage. They will learn these at Level 4.) How does the shape change when you change the lengths of the sides? (Changing all the side lengths but leaving the angles the same will make a larger version of the same shape. Changing the lengths of just one, two or three sides will change the angles and therefore change the shape.)*

Support
Some children may still need to hold their right angles against the corner of a shape in order to establish whether it is greater than, less than or equal to a right angle. Revise doing this with simple, familiar polygons – triangles and quadrilaterals – before the children try the questions on **Pupil Book 3 page 35**.

Answers for Pupil Book 3 page 35
1 a B, F, G, J, L, N   b K, M
   c A, C, D, E, H, I
2 a A < 90°, B > 90°, C 90°, D 90°, E > 90°
   b F < 90°, G > 90°, H 90°, I 90°
   c J 90°, K < 90°, L > 90°, M 90°
   d N 90°, O < 90°, P > 90°, Q 90°, R 90°, S 90°
3 a 90°   b < 90°   c > 90°   d < 90°

Sketch 2D shapes

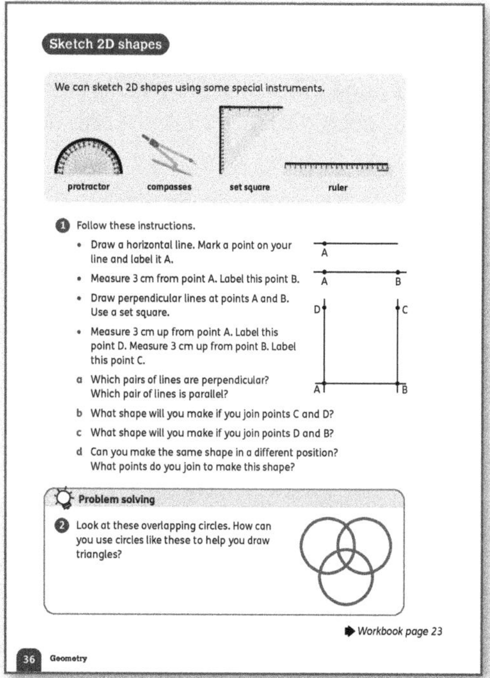

Materials
Rulers; set squares; pairs of compasses and protractors

Warm-up
- Make sure the children understand the terms *horizontal*, *vertical*, *perpendicular* and *parallel*.
- Show the mathematical instruments you have brought and look at the pictures on **Pupil Book 3 page 36**. Discuss how we use each instrument:
  - *We use a ruler for measuring and to draw lines of a given length.*
  - *A set square helps us to draw right angles.*
  - *We use a protractor to measure angles in degrees and to draw angles.* Demonstrate how to line up the protractor correctly to draw an angle.
  - *A pair of compasses helps us to draw curves and circles.* Demonstrate how to use a pair of compasses to draw a curve and then a complete circle.

(Note that the children do not need to work with protractors and compasses at this stage – you can omit these if you feel they are too advanced for your class.)

Focus
- Show the children, step-by-step, how to begin sketching and constructing 2D shapes.
- You can follow the steps in question 1 on **Pupil Book 3 page 36**, or work more slowly through these skills, allowing the children to draw their own constructions. The important steps they need to understand are:
  - drawing a line using a ruler
  - distinguishing between horizontal and vertical lines
  - marking a point and lining up the 0 mark on the ruler on the point
  - marking a point a given distance away
  - constructing a perpendicular line
  - joining points to create shapes.

- <u>Problem solving</u>: In question 2, the children explore how three overlapping circles can help when constructing triangles.

Follow-up
Use **Workbook 3 page 23** to give the children extra practice in sketching 2D shapes with given lengths and angles.

Challenge
The children can try to construct shapes such as a regular hexagon and a regular pentagon. If you have pairs of compasses available, let them explore creating overlapping circles and joining the intersections.

Support
If children struggle with constructions, start with these simple constructions:
- lines of a given length
- squares with sides a given length
- rectangles.

Answers for Pupil Book 3 page 36

1 a perpendicular: AB and AD, AB and BC
parallel: AD and BC
  b a square   **c** a triangle
  d Join A to C. (Alternatively, join C to D and B to D *or* join C to D and A to C.)
2 Draw straight lines between three points where the circles intersect (cross each other).

Answers for Workbook 3 page 23

1 Individual answers

Symmetry

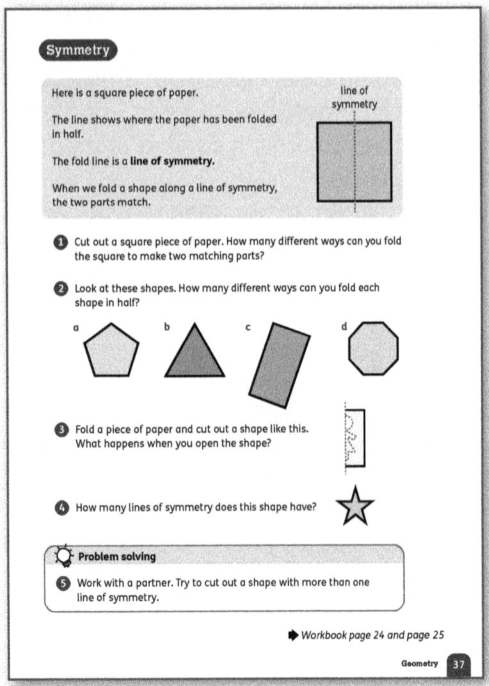

Materials
Sheets of paper and card, scissors; pinboards or dotty paper, mirrors

Warm-up
- Introduce the concept of *symmetry* by pointing out things in the classroom that have *lines of symmetry*.
- Demonstrate folding a rectangular sheet of paper in half along its length, so that one half fits exactly on top of the other – the fold line is a line of symmetry. Repeat this by folding the paper along its width.
- Show the children that folding the sheet along a diagonal produces two halves that are also equal in size, but this fold line is not a line of symmetry because one half does not fit exactly on top of the other.

Focus
- Ask the children to cut out shapes from paper or card. They can explore whether each shape has any lines of symmetry by folding the shape to see whether one half fits exactly on top of the other.
- The children work in pairs. One child constructs half a shape on a pinboard (or dotty paper) and the other child completes the shape so that it has a line of symmetry. The children could also use a mirror to help them draw shapes with one line of symmetry.
- Let the children complete questions 1–3 on **Pupil Book 3 page 37** to make sure they understand the concept of symmetry.
- In question 4, scaffolding is reduced. Some children may be able to draw a star and work out the lines of symmetry but others may need to cut out a star and find the lines of symmetry by folding.
- <u>Problem solving</u>: In question 5, the children can work in pairs to try to draw and cut out a shape with more than one line of symmetry.

Follow-up
- Use **Workbook 3 page 24** to assess whether the children can identify *symmetrical* shapes and draw a line of symmetry correctly.
- Use **Workbook 3 page 25** to teach the children to complete symmetrical shapes and patterns.
- Ask the children to cut out multiple shapes such as chains of paper dolls and to try other paper crafts.

Challenge
Give the children this challenge: *Draw a line. Draw a shape on one side of the line to make one half of a symmetrical shape. Swap with a partner. Complete each other's shapes.*

Support
Draw some isosceles triangles and rectangles and let the children draw the line of symmetry on each. Then gradually move on to more complex shapes such as butterflies.

Answers for Pupil Book 3 page 37

1 4
2 a 5   **b** 3   **c** 2   **d** 8
3 3 It makes the shape of a person or doll.
4 5
5 Individual answers/Practical work

Answers for Workbook 3 page 24

1. Ticked: A, C, D, E, H, I, M, O, T, U, V, W, X, Y
   One vertical line of symmetry drawn down the centre of: A, M, T, U, V, W, Y
   One horizontal line of symmetry drawn across the centre of: C, D, E
   Two lines of symmetry (one vertical, one horizontal) drawn on: H, X

2. Individual answers. For example: DEED (horizontal line of symmetry), MOM (vertical line of symmetry)

Answers for Workbook 3 page 25

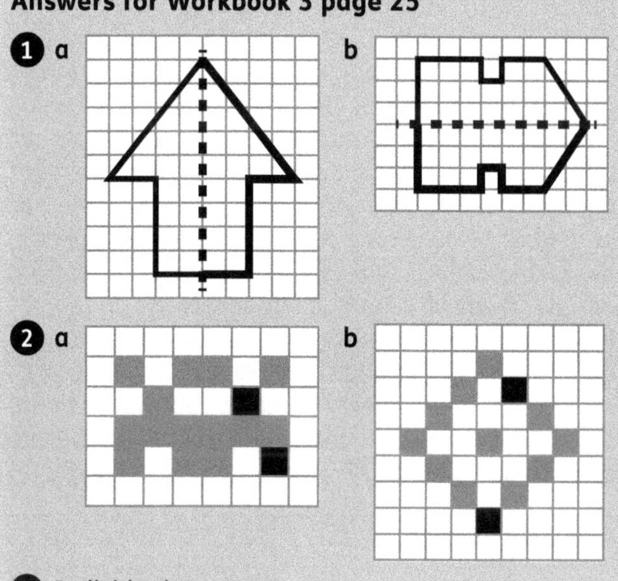

1. a   b

2. a   b

3. Individual answers

End-of-unit check

Check the children's understanding of this unit by asking these questions:
- *What is a polygon?* (a shape with straight sides)
- *How do you know if a shape is not a polygon?* (All its sides are not straight. For example, a circle and a semicircle are not polygons).
- Provide some sets of mixed polygons. *Put these polygons in order from least to greatest number of sides.*
- *Which of these polygons are regular/irregular?*
- Show a polygon and point to each angle in turn. *Is this angle less than, equal to or greater than a right angle?*
- *Draw four different-shaped irregular pentagons.* (Possible drawings: a house shape, a 'squashed' pentagon shape, a pentagon with two acute angles)
- *Draw sketches of a triangle, a rectangle and a square.*
- *Show me how to use folded paper to cut out symmetrical shapes.*
- Give the children half a shape drawn with a mirror line. *Complete this drawing to make a symmetrical shape.*

Mixed practice 1

You can use Mixed practice 1 on **Pupil Book 3 pages 38–39** to assess the children's confidence in the concepts from Units 1–6.

Answers for Mixed practice 1, Pupil Book 3 pages 38–39

1. Possible answers:
   Mathematics is used all the time in real life, for example when shopping or measuring. If you are the fastest but get lots of answers wrong, then you are not the best at maths. Making mistakes in maths is very important because our brains grow fastest when we learn from our mistakes.

2. a 10   b 10

3. 390, 391, 392, 393, 394, 395, 396

4. Many answers are possible. For example: 158, 815, 518, 581, 18, 81, 85, 15
   In order: 15, 18, 81, 85, 158, 518, 581, 815

5. Many answers are possible. For example, partition 749 as: 700 and 49; 700, 40 and 9; 600, 40 and 109

6. a 450   b 380   c 610   d 840

7. a 200   b 300   c 400   d 1000

8. a 7 cm   b 4 cm

9. Individual answers (The length of the third side will depend on the angle between the 5-cm and 3-cm lines.)

10. 
    1, 5, 9, 13

11. a Individual answers. For example:

    b Possible answers: Start at 2 and count on in 2s.
    Start at 2 and count the even numbers.
    Start at 2 and add 2 each time.

12. a a right angle   b 4

13. a D, F   b C, G   c A, B, E
    d B, D, G   e A, E, D, F   f A, C, E, F

14. Possible answers: Parallel lines stay the same distance from each other and never meet. Perpendicular lines meet at a right angle.

    Parallel lines   Perpendicular lines

Addition and subtraction

Learning objectives
- Recall addition and subtraction facts that bridge 10.
- Count on and back in steps of constant size.
- Understand and use commutative and associative properties of addition.
- Understand the inverse relationship between addition and subtraction, and how both relate to the part–part–whole structure.
- Estimate, add and subtract whole numbers with up to 3 digits (regrouping of ones or tens).
- Estimate the answer to a calculation and use inverse operations to check answers.
- Add and subtract numbers mentally, including: a 3-digit number and ones; a 3-digit number and tens; a 3-digit number and hundreds.
- Add and subtract numbers with up to 3 digits using columns.
- Regroup 3-digit numbers, using hundreds, tens and ones.
- Apply place-value knowledge to known additive number facts (scaling facts by 10).
- Recognise and calculate complements to 100 and complements of multiples of 10 or 100 (up to 1000).
- Solve problems, including missing number problems and more complex addition and subtraction.

Key words
addition subtraction strategy add subtract count on count back total sum difference fact family hundreds tens ones regroup column estimate inverse operation carry

Unit introduction

Materials
Number strips, number lines and number tracks; 100 charts; a set of number facts cards, as shown in the Teaching guidance below; two specially marked cubes (mark these yourself with stickers: one with 4, 5, 6, 7, 8, 9; the other with 5, 6, 7, 8, 9, 10); local coins (or copies of coins drawn and cut out from card)

Teaching guidance
Spend some time reviewing the *addition* and *subtraction* facts that the children are expected to know at this stage and remind them of the *strategies* they can use for *adding* and *subtracting* numbers mentally. They should know:
- doubles facts to 20 (1 + 1, 2 + 2, etc.)
- plus 1 facts (*counting on* to find the number 'after')
- facts for 'near doubles' (numbers that are 1 apart), so for 5 + 6 they work out double 5 plus 1
- plus 2 facts (skip counting in twos, for example 12 + 2 = 14)
- pairs (bonds) that make 10 and 20.

Here are some ideas for reviewing addition and subtraction facts:
- Use number strips, number lines and number tracks to assist with adding and subtracting.
- Revise the use of the 100 chart as a tool and method for carrying out addition and subtraction. Remind the children that each step down the grid adds 10 and each step to the right adds 1. Show them how to add two numbers by moving down and right across the grid. Also remind them that each step up the number grid takes 10 away and each step to the left takes 1 away. Show them how to subtract one number from another by moving up and left across the grid.
- Play a memory game to reinforce number facts and equivalent forms of the same calculation. Make a set of cards with simple additions, subtractions and totals on them. For example, these cards can be used to revise facts to 10:

4 + 3	3 + 4	10 – 3	7	8 – 5	4 + 4	10 – 3
0	1	3	5	7	8	10
5 + 5	6 + 4	9 – 2	10 – 5	9 – 9	10 + 0	8 – 2
7 – 0	5 – 4	6 + 4	4 + 6	3 + 7	7 + 3	6 + 0

Cut out the cards and shuffle them. The children play in small groups. The cards are laid out face down and the children take turns to turn over two cards. If they are equivalent, the child can remove them and score a point. If not, they turn the cards back over and the next child takes a turn. Continue until all the cards have been turned over. Adapt the activity for numbers to 20 and groups of three or four small numbers.
- Use two specially marked cubes (marked 4–9 and 5–10) to play addition games for numbers to 20. The children throw the cubes and find the *total* as quickly as they can. You can adapt this game to include subtraction by marking one cube with the numbers 0–5 and the other with 15–20 (or a similar range). Subtraction games must include the instruction to subtract the smaller number from the greater at this stage.
- If it is appropriate, use local coins to make amounts to 20. Give an amount, such as 5 (cents), and ask: *How many more (cents) do you need to make 20?* Use questions such as: *I had 20c. I dropped some coins and now I only have 13c. How much did I drop?* Money is also a suitable context for teaching and reinforcing pairs of multiples of 5 that make 100.

Number facts and families

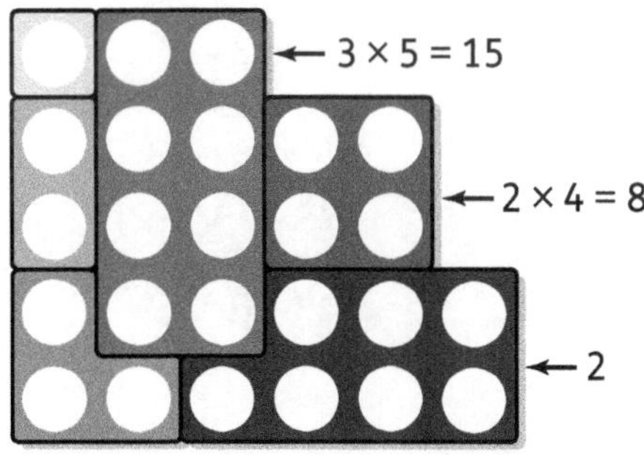

Materials

Numicon number frames (pages 21–22); ten frames (page 21) and counters

Warm-up

- <u>Think and share:</u> Turn to **Pupil Book 3 page 40** and discuss the picture of a group of Numicon number frames with the class.
- Let the children work in pairs to talk about the best way to work out what the whole picture represents.
- When you ask them to share their strategies with the class, focus less on 'who got the right answer' than on 'how did you work it out?' Let each pair explain their working, and other pairs can share if they did it differently. For example:
  - 'We counted up the columns of 5 dots: 5, 10, 15. Then we counted columns of 4 dots: 4, 8. There is one column of 2 on the right. 15 + 8 + 2 = 25.' (This may also be seen as arrays.)

- Other children may add up the numbers in other ways.

Focus

- Discuss questions 1 and 2 with the class and let the children work through them in pairs.
- For question 1a, the children can use number frames to model their calculations.

- Note that there is no need for the children to write all the scaled-up *fact families* in question 2. Focus on verbal and mental working.

Challenge

Children who are very confident scaling by 10 or 100 can make the holes in the frames (or the counters in the ten frames) represent other numbers, such as 5 or 3. Give the children some numbers to represent using the number frames or counters.

Support

- By now most children should be familiar with number complements to 10. However, they may need reinforcement of counting in *tens* and adding multiples of 10 to make 100. You could use ten frames and counters labelled 10 (make these from card if necessary). Ensure that the children are fluent in working with multiples of 10 up to 100 before moving on to working with *hundreds*.
- 'Target games' (page 30) is useful for making number complements to 10, to 100 and to 1000.

Answers for Pupil Book 3 page 40

<u>Think and share:</u> Individual answers (see the lesson notes above)

The frames represent a total of 25.

When each hole represents 10, the number frames represent a total of 250.

1 a Possible answer: Number frame showing 9 and 1

$9 + 1 = 10$, $1 + 9 = 10$, $10 - 1 = 9$, $10 - 9 = 1$
(or the fact family for any other bonds to 10)

b Possible answer: We can use our knowledge of fact families to check inverse operations. For example, we can check $10 - 8 = 2$ by using addition: $8 + 2 = 10$.

2 a

$5 + 5 = 10$	$10 - 5 = 5$	
$10 + 0 = 10$	$10 - 0 = 10$	$10 - 10 = 0$
$9 + 1 = 10$	$1 + 9 = 10$	$10 - 9 = 1$ $10 - 1 = 9$
$8 + 2 = 10$	$2 + 8 = 10$	$10 - 8 = 2$ $10 - 2 = 8$
$7 + 3 = 10$	$3 + 7 = 10$	$10 - 3 = 7$ $10 - 7 = 3$
$6 + 4 = 10$	$4 + 6 = 10$	$10 - 6 = 4$ $10 - 4 = 6$

Other answers are also possible. For example:
$3 + 1 + 4 + 2 = 10$ $10 - 3 = 1 + 4 + 2$ etc.

b The children scale up all the numbers in their fact families in 2a to make $50 + 50 = 100$, $100 - 50 = 50$, etc.

c The children scale up again to make $500 + 500 = 100$, $1000 - 500 = 500$, etc.

Addition and subtraction facts

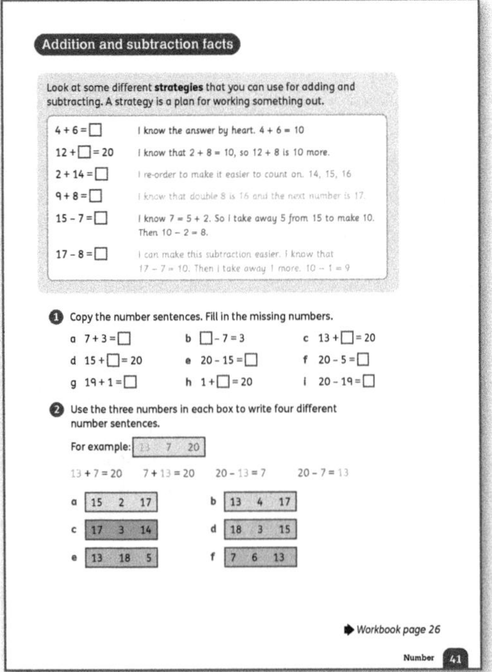

Warm-up

- Work through the examples on **Pupil Book 3 page 41**.
- Explain the word *strategies*. A strategy is something we can use or try to help us work something out. Discuss each example and the strategy it shows.
- In the first example, 4 + 6, the child knew the answer by heart. Ask: *What other additions do you know that make a total of 10?* Let the children name known number facts. Ask: *What could you do to find the total if you don't know the addition?* (For example, they could count on fingers, count on, put counters together).
- Next look at 12 + ☐ = 20. Say: *'I know that 2 + 8 = 10'* – how does this help us? Demonstrate using partitioning: 12 = 10 + 2, 20 = 10 + 10.
  So, the question is, what must we add to 10 + 2 to make it equal to 10 + 10?
- The third example shows counting on, which most children should be familiar with. The next example uses near doubles. Ask: *What other doubles do you know?* Revise doubles up to double 10, and then double various multiples of 10: *If double 2 is 4, what is double 20? If double 3 is 6, what is double 30?*
- Continue working through the examples in this way, expanding on each strategy to explore how it can be used for a range of calculations.

Focus

- Once you have revised the basic number facts, let the children work on their own to complete questions 1 and 2 on **Pupil Book 3 page 41**.
- Encourage the children to try the strategies you have been reviewing and to talk about how they are working out each calculation.
- In question 1c encourage the children to make the connection between addition and subtraction and to develop checking strategies using inverse operations.

- In question 2, the children use each set of three numbers to make two addition and two subtraction sentences. They may gradually make some observations:
  - The two smaller numbers add up to the greater number.
  - The two smaller numbers can be added up in any order (commutative property of addition).
  - Either of the smaller numbers can be subtracted from the greatest number to give the difference.

Follow-up

Use **Workbook 3 page 26** to consolidate addition and subtraction of 1- and 2-digit numbers. Make sure the children understand that they apply the operation in the box to each of the numbers on the left (top, middle and bottom). The outputs on the right are given in the same order.

Interesting mistakes

When looking at calculations such as 12 + ☐ = 20, children may say: '2 + 8 = 10, so 12 + 18 = 20'. Let them examine and explain what the mistake is here. Demonstrate counting on from 12 to 20 on a number line or 100 chart, as well as partitioning in order to show the addition.

When using 'near doubles', for example in 9 + 8 = ☐, children may rush to add or take away a 1, instead of thinking about which is needed (for example, '9 + 9 = 18, so 9 + 8 = 19'). Let them examine the mistake and explain what happened and how to correct it.

Answers for Pupil Book 3 page 41

1 a 10   b 10   c 7   d 5   e 5
   f 15   g 20   h 19   i 1

2 a 15 + 2 = 17   2 + 15 = 17
      17 – 15 = 2   17 – 2 = 15
   b 13 + 4 = 17   4 + 13 = 17
      17 – 13 = 4   17 – 4 = 13
   c 3 + 14 = 17   14 + 3 = 17
      17 – 3 = 14   17 – 14 = 3
   d 3 + 15 = 18   15 + 3 = 18
      18 – 3 = 15   18 – 15 = 3
   e 13 + 5 = 18   5 + 13 = 18
      18 – 5 = 13   18 – 13 = 5
   f 7 + 6 = 13   6 + 7 = 13
      13 – 7 = 6   13 – 6 = 7

Answers for Workbook 3 page 26

1 a 12, 13, 17   b 2, 13, 37
   c + 5   d – 5
   e + 8   f 0, 7, 19
   g 8, 21, 25   h 16, 12, 20

2 Individual answers

Addition and subtraction problems

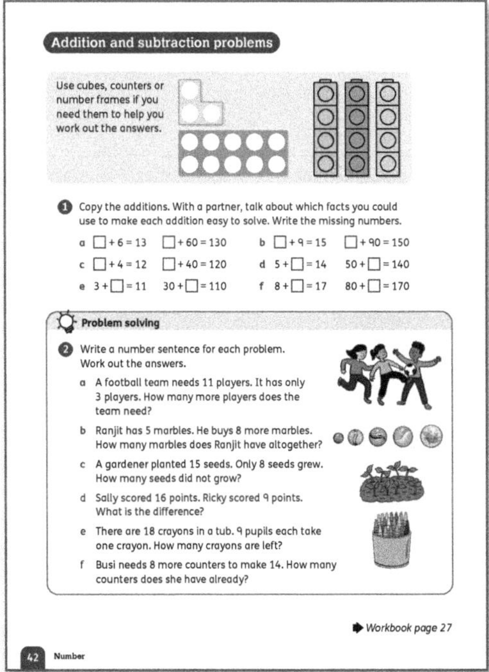

Materials
Interlocking cubes (page 21) or counters; Numicon number frames (pages 21–22); number tracks and number lines (page 27)

Warm-up
Play 'Target number' (page 26) to practise addition facts to numbers between 10 and 20.

Focus
- Let the children work in pairs or small groups to complete **Pupil Book 3 page 42** question 1.
- The point of these questions is not so much to focus on the answers (the children can very easily *count on* or *count back* to find these), but for the children to use number facts such as bridging across 10 and using known doubles.
- Encourage the children to explain their thinking and discuss what number facts they might use to work out each answer.
- There are many ways of working each one out. You might use cubes or counters and *regroup* them in order to explore the number facts. Working with counters or cubes also reinforces the children's understanding of the conservation of number (if we make the total, we can rearrange the same number of counters in different ways without altering the total).
- <u>Problem solving:</u> Read through question 2 with the class if necessary and make sure they understand the word problems. If you wish, you can work through some or all of them together.
- Again, there are different ways to work out the answers. The children may use number tracks, number lines, counters, and so on. They may count on or count back, or they may use number facts.

Follow-up
Use **Workbook 3 page 27** to consolidate complements to 100.

Challenge
Use the ideas from 'True or false/Explain the mistake' (page 29) to challenge the children. For example, they can explain the mistakes in calculations such as:
$17 - 3 = 20 \qquad 25 - 4 = 11 \qquad 300 - 20 = 100$.

When we ask children to work out what the mistake is, they need to examine the properties of the numbers they are working with, and think of the kinds of strategies they might use. These are also open-ended questions. Different children might correct the mistakes in different ways. For example, different possible correct versions of $17 - 3 = 20$ will include:
$17 + 3 = 20$, $17 - 3 = 14$ and $23 - 3 = 20$.

You can also use 'Magic squares' (page 29) as additional challenge material.

Support
Use Numicon number frames or counters to reinforce number facts to 10, then to 20. Give the children practice in doubling using concrete materials. Give addition and subtraction questions related to concrete objects, for example: *I have 18 crayons. I take away 5. How many are left?*

Interesting mistakes
It is useful to examine children's miscalculations and work out 'what happened'. This will vary from mistake to mistake, but in every case it is useful to identify the precise location of the mistake – for example an incorrect addition, adding a ten instead of a one, adding a one instead of subtracting.

Answers for Pupil Book 3 page 42

① a 7, 70	b 6, 60	c 8, 80
d 9, 90	e 8, 80	f 9, 90
② a $11 - 3 = 8$	b $5 + 8 = 13$	c $15 - 8 = 7$
d $16 - 9 = 7$	e $18 - 9 = 9$	f $14 - 8 = 6$

Answers for Workbook 3 page 27

① a 90	b 75	c 60	d 65
e 50	f 35	g 70	h 55
i 20	j 5		
② a 10	b 85	c 20	d 75
e 30	f 65	g 40	h 55
i 50	j 35		

③ There are two pairs each of: 0 and 100, 10 and 90, 20 and 80, 30 and 70, 40 and 60
There is one pair of: 50 and 50

④ a 500 steps	b 900 steps	c 1000 steps
d 500 steps	e 300 steps	f 300 steps

Addition and subtraction patterns

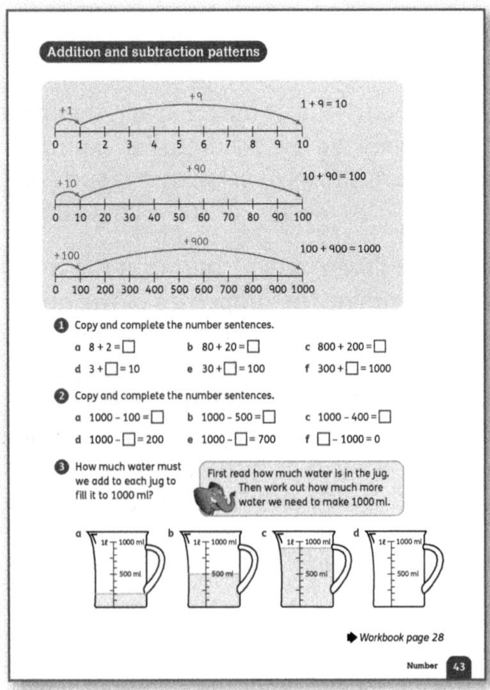

Warm-up
- Work through the examples at the top of **Pupil Book 3 page 43** with the class.
- Point out that patterns are important, and that recognising them can make it much easier to add and subtract numbers.
- You may also want to point out that addition to 100 and 1000 is particularly useful in the context of measurements before the children work through question 3.

Focus
- In question 1 read through each row of questions and draw the children's attention to what they notice about the questions (scaling up to tens, then to hundreds).
- Count in hundreds to 1000 and backwards from 1000 to 0 before the children complete question 2.
- In question 3 ask questions about the jugs, such as:
  - *How much does a full jug hold?* (1000 ml)
  - *How much would you need to fill it from empty?* (1000 ml)
  - *What if it is half full?* (500 ml)
  - *What do the little lines on the jug show us?* (100 ml and 250 ml)

Follow-up
Use **Workbook 3 page 28** to consolidate the children's understanding of breaking down and building up numbers using addition and subtraction facts. This will also help to prepare them for adding and subtracting in *columns*, which is coming up in the next lesson.

Challenge
Before the children go on to add in columns, you may wish to use number lines for addition using multiples of 5. For example, you could work through some different ways to make 100 using a number line:

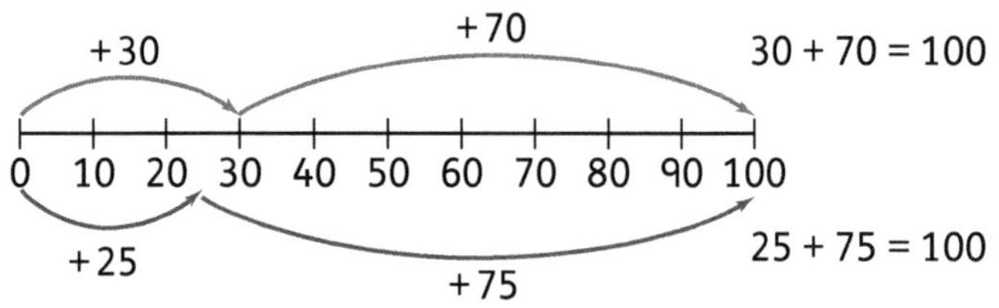

Using a number line and counting on	Reordering and using known facts
25 + 75 Start with the greater number. Count on 5. Count on the rest in tens.	25 + 75 = 25 + 5 + 70 = 30 + 70 = 100

Once you have worked through this example, give the children questions based on counting forwards in multiples of 25. Here are some ideas:

1 Copy the number sentences. Write the missing numbers.
   a $\square + 25 = 100$ **b** $\square + 60 = 100$
   c $\square + 95 = 100$ **d** $30 + \square = 100$
   e $45 + \square = 100$ **f** $20 + \square = 100$

2 Nisha joined two pieces of ribbon to make a piece 100 cm long. Work out the length of the second piece of ribbon.

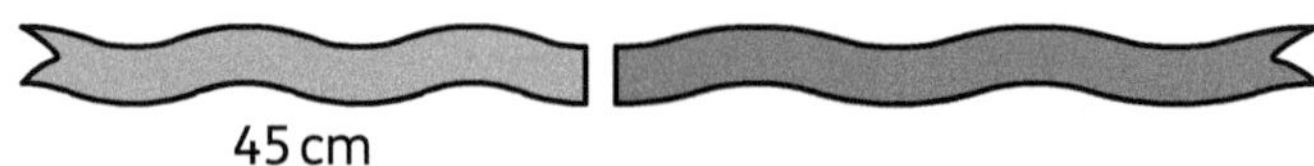

45 cm

3 Fill in the missing numbers to make these number sentences true.
   a $35 + \square = 95 + 5$ **b** $50 + 50 = 45 + \square$
   c $20 + 80 = 85 + \square$ **d** $75 + \square = 90 + 10$

Support
Use 'Number meet-up' (page 26) and 'Counting on a number line' (page 27) to reinforce the work in this lesson.

Answers for Pupil Book 3 page 43

1 a 10 b 100 c 1000
  d 7 e 70 f 700

2 a 900 b 500 c 600
  d 800 e 300 f 1000

3 a 800 ml b 500 ml
  c 100 ml d 1000 ml

Answers for Workbook 3 page 28

1 a 90 b 50 c 70 d 80

2 a 15 b 75 c 65
  d 75 e 15 f 50

3 a 58 b 33 c 47
  d 81 e 26 f 19

4 49 + 51   89 + 11   33 + 67

Work in columns

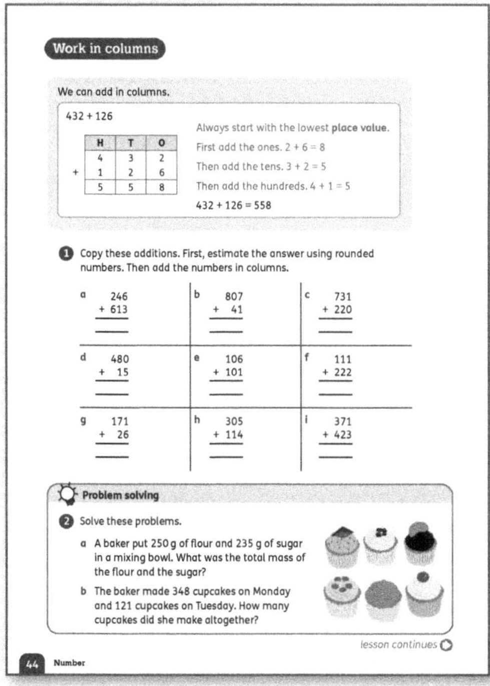

Work in columns

We can add in columns.

432 + 126

H	T	O
4	3	2
1	2	6
5	5	8

Always start with the lowest **place value**.
First add the ones. 2 + 6 = 8
Then add the tens. 3 + 2 = 5
Then add the hundreds. 4 + 1 = 5
432 + 126 = 558

1 Copy these additions. First, estimate the answer using rounded numbers. Then add the numbers in columns.

a 246 + 613
b 807 + 41
c 731 + 220
d 480 + 15
e 106 + 101
f 111 + 222
g 171 + 26
h 305 + 114
i 371 + 423

Problem solving

2 Solve these problems.

a A baker put 250 g of flour and 235 g of sugar in a mixing bowl. What was the total mass of the flour and the sugar?

b The baker made 348 cupcakes on Monday and 121 cupcakes on Tuesday. How many cupcakes did she make altogether?

lesson continues

44 Number

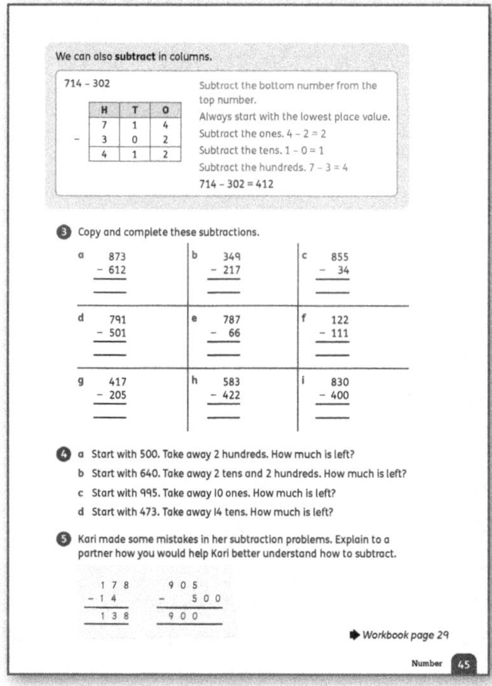

We can also **subtract** in columns.

714 − 302

H	T	O
7	1	4
3	0	2
4	1	2

Subtract the bottom number from the top number.
Always start with the lowest place value.
Subtract the ones. 4 − 2 = 2
Subtract the tens. 1 − 0 = 1
Subtract the hundreds. 7 − 3 = 4
714 − 302 = 412

3 Copy and complete these subtractions.

a 873 − 612
b 349 − 217
c 855 − 34
d 791 − 501
e 787 − 66
f 122 − 111
g 417 − 205
h 583 − 422
i 830 − 400

4
a Start with 500. Take away 2 hundreds. How much is left?
b Start with 640. Take away 2 tens and 2 hundreds. How much is left?
c Start with 995. Take away 10 ones. How much is left?
d Start with 473. Take away 14 tens. How much is left?

5 Kari made some mistakes in her subtraction problems. Explain to a partner how you would help Kari better understand how to subtract.

178 − 14 = 138   905 − 500 = 900

Workbook page 29

Number 45

Materials

Base-ten blocks (100-flats, 10-rods and ones cubes) (page 22); place-value tables (page 23); squared paper

Warm-up

- Introduce column addition using base-ten blocks or illustrations. Whichever materials you use, it is very important to relate your working to the partitioning of numbers into hundreds, tens and *ones*.
- You may want to start with some 2-digit examples, such as the addition shown top right.

Hundreds	Tens	Ones
	(4 ten-rods)	(3 ones)
	(6 ten-rods)	(6 ones)

43
+ 26
———

- Make sure you use language precisely:
  - *3 ones plus 6 ones is equal to 9 ones.*
  - *4 tens plus 2 tens is equal to 6 tens.*
- Demonstrate how to set out addition in columns using only the digits, alongside the concrete representation.
- Go through several examples with 2-digit numbers.

Focus

- Work through some column additions using 3-digit numbers, in the same way. Start with additions that do not involve regrouping. For example:

Hundreds	Tens	Ones
(2 hundred-flats)	(1 ten-rod)	(5 ones)
+ (1 hundred-flat)	(2 ten-rods)	(3 ones)

215
+ 123
———

- Then move on to additions that involve regrouping. For example:

Hundreds	Tens	Ones
(2 hundred-flats)	(4 ten-rods)	(6 ones)
+ (1 hundred-flat)	(7 ten-rods)	(3 ones)

246
+ 173
———

- You may need to work with column addition and subtraction over a few lessons. Work with concrete materials and give the children plenty of practice in representing the numbers and then working in each column to find the total.
- Once the children begin to write the digits in columns, you will need to emphasise the importance of:
  - lining up the digits correctly (using squared paper to help, if necessary)
  - working from right to left (ones, then tens, then hundreds).
- Note that the concrete representations should only be used initially, in order to support the children's understanding. Once the children understand the column algorithm (how to work in columns using only the numbers), they can begin to use the number facts they have practised in earlier lessons.

- As the children work through **Pupil Book 3 page 44** and **page 45**, ensure that they show how they work the answers out. Children of this age sometimes start to think that the most important thing is just to give an answer. It is important that they understand that showing how they work something out is more important than getting the correct answer.
- In question 1 remind the children of the importance of *estimating* the answer using rounded numbers so they can see if their solution is sensible.
- <u>Problem solving:</u> Question 2 presents two addition word problems. The children could use column addition or other strategies.
- Before the children work on question 3, use place-value tables and base-ten blocks to model subtraction without regrouping.
- In question 3 the children use column subtraction where no regrouping is needed. Some children may need support in parts c and e to realise that the empty columns represent 0. If time allows, encourage the children to check their answers by adding.
- Question 4 involves solving subtraction word problems. Some children may benefit from using a place-value table and/or base-ten blocks.
- Question 5 requires the children to recognise positioning mistakes in columnar subtraction. Some may benefit from using base-ten blocks and/or a place-value grid to help them understand the mistakes.

Follow-up

On **Workbook 3 page 29**, the children develop their column addition skills by adding three numbers (that do not involve regrouping). Keep emphasising that the children should look out for easy ways to combine numbers in tens, which will help them add efficiently.

Challenge

Give the children addition and subtraction questions that do not require regrouping and ask them to work these out mentally. You can also ask the children to make up their own missing-digit addition or subtraction questions using columns, similar to those in question 3 on the Workbook page. They must check that the question can be solved before they give it to a partner to work out.

Support

Some children may need to work slowly, step by step, to grasp the idea of adding or subtracting in columns. For these children, you may need to use several strategies:
- Go back to addition of 2-digit numbers (numbers with tens and ones).
- Use place-value tables and demonstrate using cubes to represent tens and ones.
- Go back to expanding numbers (expressing them as hundreds, tens and ones). For example:
  58 + 11
  = 5 tens 8 ones + 1 ten 1 one
  Add the ones. 8 + 1 = 9
  Add the tens. 50 + 10 = 60
  6 tens 9 ones = 69.

Once the children are comfortable working with 2-digit addition, move on to adding a 1-digit number to a 3-digit number (for example, 512 + 4). Then move on to 3-digit plus 2-digit numbers (for example, 135 + 12) and finally to 3-digit plus 3-digit numbers. Remember to keep relating questions to real life, such as:
- *314 people visit a shop in the morning, and another 143 visit in the afternoon. How many people visit the shop altogether? (457)*
- *I have $215 and I get $111 more. How much do I have altogether? ($236)*

Work in the same way with subtraction.

Interesting mistakes

The children are likely to make a variety of interesting mistakes at this level, including errors in lining up the digits of each place value correctly. Whenever calculation errors occur, ask questions to explore them, for example:
- *Let's look at each number. Where are the ones? How many ones are there in this number? Where are the tens? How many tens are there in this number? Where are the hundreds? How many hundreds are there in this number? Are they lined up?*
- In the case of a story problem, ask questions such as:
  - *Are we adding or subtracting? How do you know?*
  - *What are we trying to find out?*
  - *Where do you think the mistake is? How can we check?*

Answers for Pupil Book 3 pages 44–45

1 a 859   b 848   c 951
   d 495   e 207   f 333
   g 197   h 419   i 794

2 a 485 g   b 469

3 a 261   b 132   c 821
   d 290   e 721   f 11
   g 212   h 161   i 430

4 a 300   b 420
   c 895   d 333

5 Possible answer: Line up the numbers so that the ones, tens and hundreds are all in the same column. (Correct answers: 164 and 405)

Answers for Workbook 3 page 29

1 a 37   b 55   c 69
   d 76   e 99   f 79

2 a 425   b 738   c 799

3 a 4   b top row 7, second row 5
   c top row 0, second row 3

4 427

5

H	T	O
	2	6	5
+	1	2	2
	3	8	7

Add or subtract multiples of 10 and 100

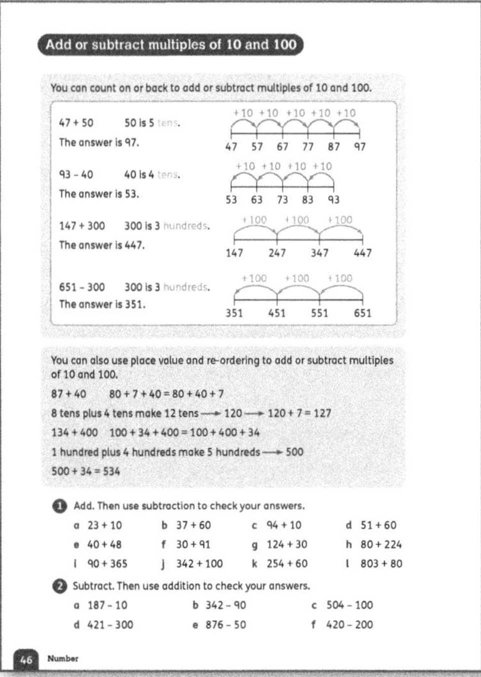

- Work through as many examples as necessary to make sure the children understand this important concept. Repeat for subtraction.
- Use place-value cards to demonstrate how to add numbers by:
  - partitioning the numbers using place value
  - then adding the hundreds, tens and ones
  - recombining the hundreds, tens and ones to find the total.
- This is an absolutely fundamental skill that the children must understand if they are to succeed in calculating. For example, you could show the class how to add 38 to 141 like this:

$$38 = 30 + 8$$
$$\underline{141 = 100 + 40 + 1}$$
$$100 + 70 + 9$$

$$38 + 141 = 179$$

Let the children work in pairs with their own sets of place-value cards to solve a range of different calculations.

- Work through the examples on **Pupil Book 3 page 46** with the class. Allow for some discussion and demonstration of using larger jumps once the children have understood the concept.
- Note that question 1 has a lot of parts. Use your discretion to decide how many parts the children in your class need to complete.
- In question 2, the children work out the subtractions and then use addition to check their answers. Check that they understand how to do this before they complete the page.

Materials

Blank number lines; place-value cards (pages 22–23); place-value tables (page 23) and concrete materials (base-ten blocks (page 22), place-value counters, etc.)

Warm-up

- Revise place value in 2-digit and 3-digit numbers. It is essential that the children appreciate that a 2-digit number consists of a number of tens and a number of ones.
- Write some 2-digit numbers on the board and discuss the value of each digit. For example, in the number 58, the 5 represents 5 tens, which are 50 and not simply 5. Repeat for 3-digit numbers.

Focus

- Teach the class how to use a blank number line as a schematic aid for calculation. Show them how to mark only the numbers they need for the calculation. For example, to add 25 and 38, they can draw a line and mark 38. Then they can jump in groups to get to the total of 63. Show them that they can jump in different ways:
  - They could jump in tens and partition the ones to bridge 60 like this:

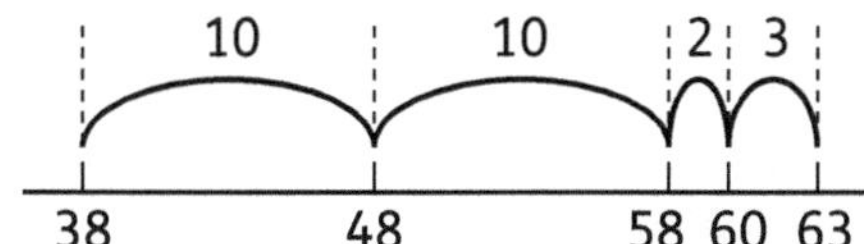

  - They could make two larger jumps like this:

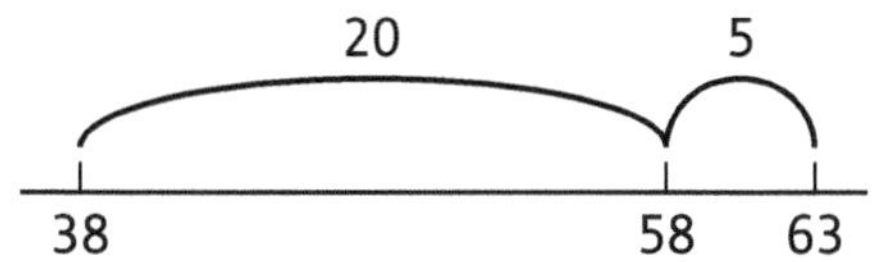

Support

Some children may need you to work through additional examples individually with them. If necessary, use place-value tables and concrete materials in addition to the number lines.

More adding and subtracting

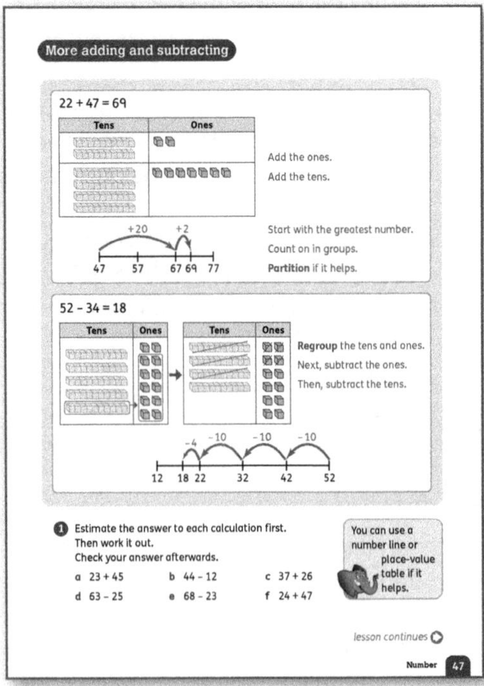

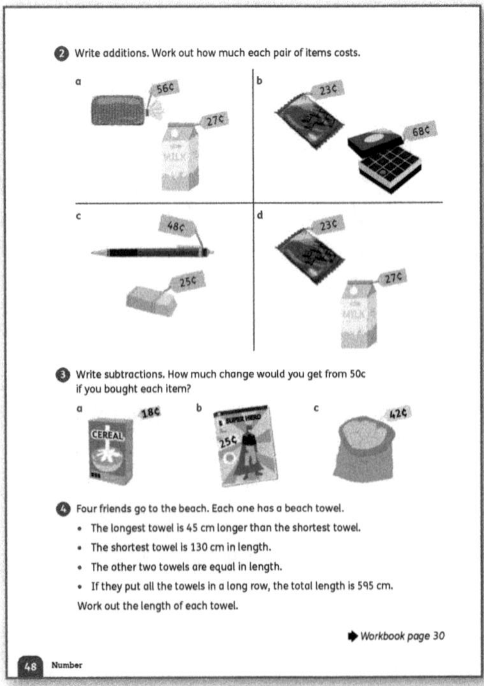

Materials

Place-value tables (page 23); base-ten blocks (page 22); place-value cards (pages 22–23) and number lines (page 27)

Warm-up

- Work through some examples of addition and subtraction using place-value tables and base-ten blocks, as shown on **Pupil Book 3 page 47**.
  In these representations, we use 10-rods in the tens column and ones cubes in the ones column. We rearrange the tens and ones to add or subtract. This is the same as adding or subtracting in columns, but using visual representations rather than digits.
- Give the children plenty of practice in adding and subtracting in this way if they need visual support.

- Note that the children will start regrouping tens to make ones, but will do this using the concrete manipulatives (base-ten blocks). They will move on to calculating in columns using regrouping of digits in the next lesson. Make sure they understand the concept of rearranging a group of ten to make ones in order to regroup before they move on to working in columns.

Focus

- It is a good idea to introduce the practice of *estimating*, then calculating and then checking using an *inverse operation*. Work through some examples with the class, for example 42 + 59.
- Ask: *Without working out the answer, do you think it will be closer to 10, 50 or 100? How can you tell which is the most reasonable estimate?*
- Let the class share their ideas. Remind them that when we work out roughly what an answer will be, we are using estimating. One way to estimate is to use rounding. So you can say: *Let's round 42 to the nearest ten. It rounds to … ? (40) What does 59 round to? (60) What do 60 + 40 add up to? (100)*
- If necessary, remind the children that 6 + 4 = 10, so 6 tens + 4 tens = 10 tens.
- Then ask individual children to come up and work out the accurate answer to 42 + 59. (101)
- Let different children demonstrate different ways we could do this, for example:
  - adding tens, adding ones, regrouping
  - adding in columns with regrouping
  - adding in jumps along a number line.
- Also demonstrate how to check the answer using subtraction. If necessary, work through one or two more examples before the children complete questions 1 to 4.

Follow-up

Let the children work through **Workbook 3 page 30**.

Support

The children may need plenty of support working with subtraction. It is not always practical to count back, and they may need you to remind them of various strategies they can try. You can ask questions such as:

- *What is ten more/ten less?*
- *Can we use partitioning?*
- *Can we use halving to help?*
- *How can we check the answer?*

The children may attempt to count on instead of recalling addition facts and using patterns to add multiples of 10 and 100. You can address this by discussing methods and by providing lots of practice with addition facts and combinations.

Encourage modelling using place-value cards and/or number lines, as well as noting down addition facts to assist, as this often makes the connection clearer for the children.

Answers for Pupil Book 3 pages 47–48

1 a $20 + 50 = 70$ 68 **b** $40 - 10 = 30$ 32
   c $40 + 30 = 70$ 63 **d** $60 - 30 = 30$ 38
   e $70 - 20 = 50$ 45 **f** $20 + 50 = 70$ 71

2 a $56 + 27 = 83c$ **b** $23 + 68 = 91c$
   c $48 + 25 = 73c$ **d** $23 + 27 = 50c$

3 a $50 - 18 = 32c$ **b** $50 - 25 = 25c$
   c $50 - 42 = 8c$

4 130 cm, 145 cm, 145 cm, 175 cm

Answers for Workbook 3 page 30

1 a 118

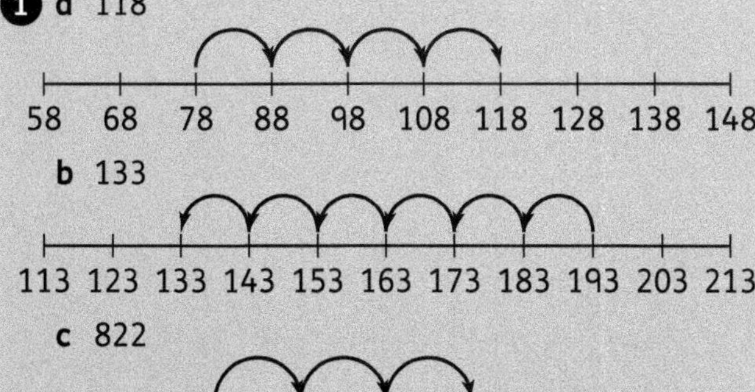

   b 133

   c 822

   d 445

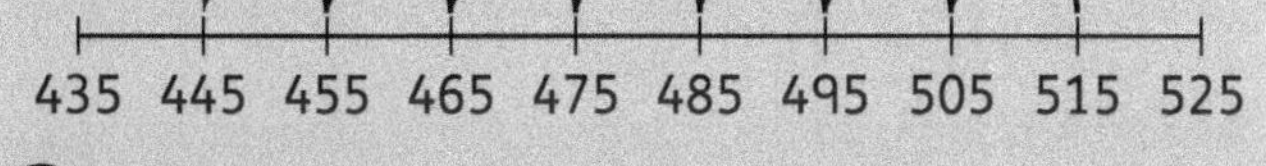

2 a

6 hundreds	3 tens	2 ones
–	2 tens	1 one
6 hundreds	1 ten	1 one = 611

(Provided as an example)

  b

5 hundreds	4 tens	8 ones
–	3 tens	5 one
5 hundreds	1 ten	3 ones = 513

  c

7 hundreds	5 tens	6 ones
+	3 tens	3 one
7 hundreds	8 tens	9 ones = 789

Addition with regrouping

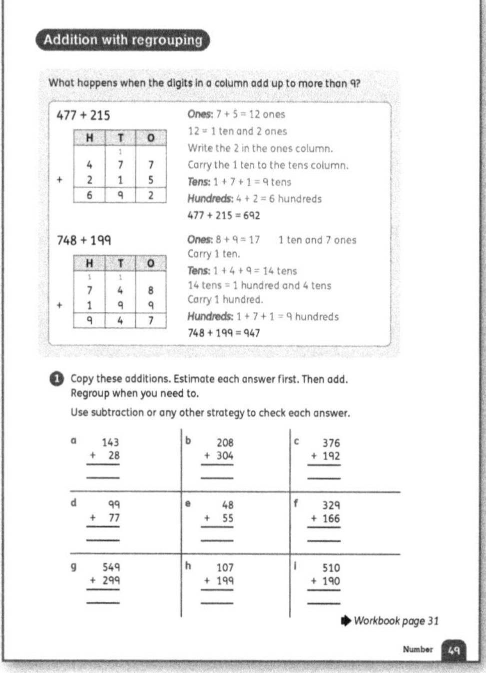

Materials

Interlocking cubes; place-value tables and base-ten blocks

When teaching the algorithm for addition in columns with regrouping, it is very important to keep drawing the children's attention back to place value, and what each number represents. If the children attempt to add in columns without an awareness of place value, they will not be able to check their answers.

As for **Pupil Book 3 page 47** and **page 48**, encourage the children to estimate a reasonable answer first, then calculate and finally check using an inverse operation or other method.

Warm-up

- Give groups of children interlocking cubes and ask them to use these to add 27 and 38 (and similar additions that involve regrouping).
- Ask them to explain what they do when they get more than 9 'ones'. (Regroup them into a ten.)

Focus

- Work step by step through the algorithm for adding in columns. Work through the examples on **Pupil Book 3 page 49**, and if needed, work through additional examples of your own. Here are some useful questions and explanations, as you work through examples of adding in columns:
  - *What are we adding to what?*
  - *How many hundreds? How many tens? How many ones?*
  - *When I write the number I am adding, I must make sure the hundreds, tens and ones are lined up. Are they lined up?*

- Once you have worked in columns and reached a total, ask the children to suggest how we can check our working: *Does it look right? How can you tell? What can we use to check?* For example, in the second example in the **Pupil Book** (748 + 199), the children may notice that 199 is just 1 less than 200, so they could check their answer by adding 200 and taking away 1.
- It is important that the children are able to add in columns, although it is not always necessary to use this method. Remind the children that they can still use their other strategies for adding and subtracting, and that using different strategies is a very useful way to check their calculations.
- Give individual children a chance to add in columns on the board too.
- The children can work independently on question 1.

Follow-up
Use **Workbook 3 page 31** to consolidate addition in columns.

Challenge
Note that for children who quickly grasp the algorithm of adding in columns, there is no benefit in them doing huge quantities of examples. The algorithm is a useful tool that is intended to help them to solve problems. Once the children have mastered the algorithm, direct them to more challenging problem-solving questions, where they might use column addition or other strategies. For example:
- *Find three 3-digit numbers that you can add to make a total of exactly 500 (or 700, or 800 or 1000).*

Support
While some children may master the algorithm quickly, others may need additional practice. Go back to addition in columns of 2-digit numbers that require regrouping the ones, for example: 73 + 18, 19 + 12.

Keep representing the tens and ones by using base-ten blocks and by expanding the numbers as you work through examples.

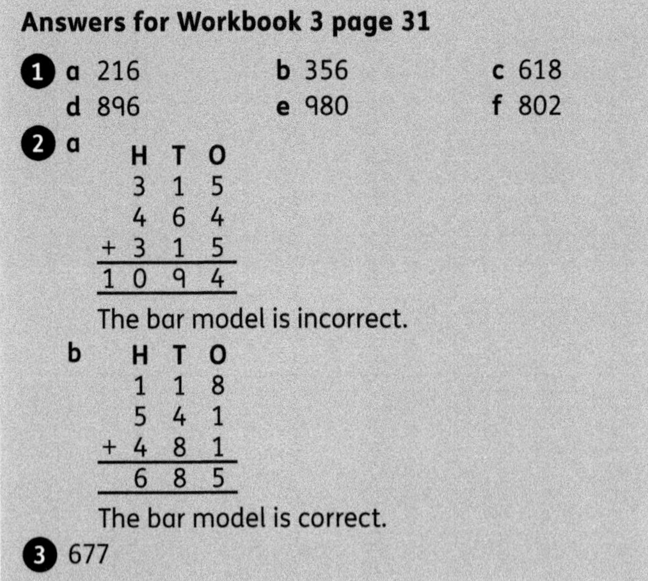

Answers for Pupil Book 3 page 49

1
a 140 + 30 = 170   171
b 200 + 300 = 500   512
c 400 + 200 = 600   568
d 100 + 80 = 180   176
e 50 + 60 = 110   103
f 300 + 200 = 500   495
g 500 + 300 = 800   848
h 100 + 200 = 300   396
i 500 + 200 = 700   700

Answers for Workbook 3 page 31

1
a 216   b 356   c 618
d 896   e 980   f 802

2 a

```
  H  T  O
  3  1  5
  4  6  4
+ 3  1  5
10  9  4
```

The bar model is incorrect.

b

```
  H  T  O
  1  1  8
  5  4  1
+ 4  8  1
  6  8  5
```

The bar model is correct.

3 677

Subtract in columns

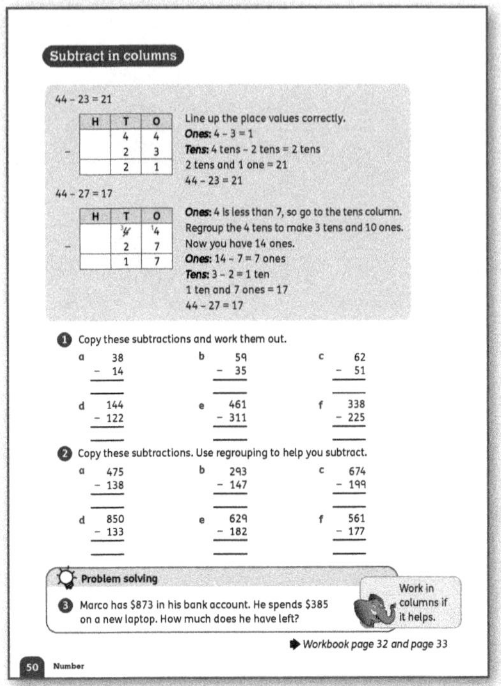

Materials
Place-value tables and base-ten blocks

Warm-up

Introduce subtraction in columns using a similar approach to the one you used for addition in columns with regrouping. Work step by step, focusing on:
- lining up the place values correctly
- working from right to left
- inspecting from the top to the digit below, and if the bottom digit is greater, regrouping from the column to the left
- representing using base-ten blocks in place-value tables
- using a variety of strategies to check answers.

Focus

- Once the children have worked through sufficient examples on the board, let them work through the questions on **Pupil Book 3 page 50**.
- In question 1, the subtractions do not involve regrouping.
- The subtractions in question 2 do require regrouping.
- Problem solving: Question 3 presents a word problem. The children could use column subtraction or another strategy.

Follow-up

Use **Workbook 3 page 32** and **page 33** to consolidate subtraction in columns.

Challenge

As for addition, there is no benefit in having the children complete numerous examples of subtraction in columns. As soon as the children have fully grasped the concepts, redirect their attention towards open-ended (or 'open middle') challenges that allow them to choose a strategy that is suited to the problem. The children should have opportunities to choose when to use their column subtraction skills and when to use other strategies.

Support

Some children may require additional practice to grasp the idea of subtraction in columns. Work through examples with 2-digit numbers, moving on to examples with 3-digit numbers, and keep reinforcing the meaning of each calculation through the use of partitioning, expanded notation and base-ten materials.

Interesting mistakes

Unlike addition in columns, which can be done either from top to bottom or from bottom to top, subtraction has to be done from top to bottom. We are always subtracting the bottom number from the top number.

Sometimes, children may get confused and want to subtract the smaller digit from the greater, irrespective of their position. In this case, use base-ten blocks and place-value tables to reinforce the sense of the calculation: *We are starting with* [top number] *and taking away* [bottom number]. *So, we need to take these ones from these ones, these tens from these tens and these hundreds from these hundreds.*

Part–part–whole

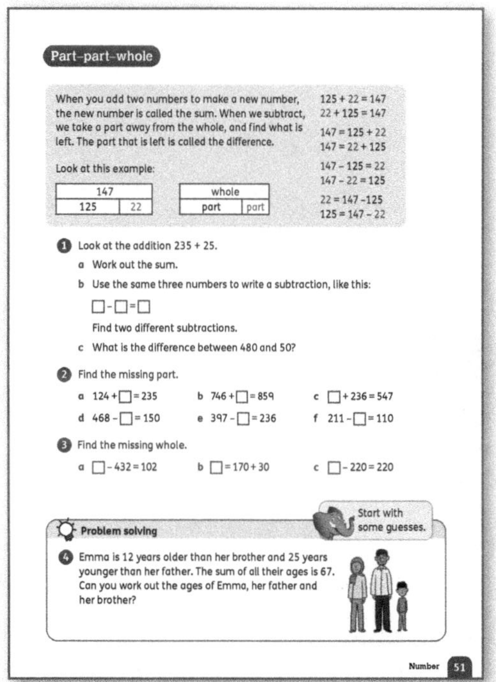

Learning that any number can be broken into parts (decomposed) and put back together (recomposed) is a key part of children's understanding of number. At Level 3, the children need to explore different ways to 'make up' and 'break up' numbers.

In our number system, it is usual for us to partition numbers into hundreds, tens and ones (and later thousands, etc.). However, we can also partition numbers in other ways. Bar models (page 22) are very useful for visualising partitioning.

Warm-up
Read through the information and examples on **Pupil Book 3 page 51**. Ask: *What different ways can we show this addition/subtraction? How does the bar model help us to see it more clearly?*

Focus
- Once you have worked through the examples, work through question 1 with the class. Then have the children work independently on questions 2 and 3.
- <u>Problem solving:</u> The children may need to use trial and improvement to find the solution to question 4.

Answers for Pupil Book 3 page 51

1 a 260
   b 260 − 235 = 25   260 − 25 = 235
   c 430

2 a 111   b 113   c 311
   d 318   e 161   f 101

3 a 534   b 200   c 440

4 Emma is 18, her father is 43 and her brother is 6.

UNIT
8 Money

Learning objectives
- Interpret money notation for currencies that use a decimal point.
- Add and subtract amounts of money to give change, using both dollars ($) and cents (c) in practical contexts.

Key words
money   cost   dollar   cent
local currency names   decimal point
placeholder   price   change

Assess the work on addition and subtraction covered in this unit by asking questions such as:
- *Which number has 2 hundreds, 8 tens and 7 ones?* (287)
- *What is 25 plus 13?* (38)
- *What is 78 plus 19?* (97)
- *What is the sum of (give two numbers)?*
- *What is the difference between (give two 2-digit numbers)?*
- *How many tens are there in a hundred?* (10)
- *What is 123 add 46?* (169)
- *What is 39 take away 16?* (23)
- *What is 60 + 30?* (90) What is 600 + 300? (900)

Give problem-solving questions that involve addition and subtraction of 3-digit numbers with and without regrouping, for example:

Some children are playing a game. There are two rounds. These are their scores:

	Jeremy	Jabu	Aareli	Wim
Round 1	170	314	459	211
Round 2	225	212	400	301

- *What is Jeremy's/Jabu's/Aareli's/Wim's total score for both rounds?* (Jeremy, 395; Jabu, 526; Aareli, 859; Wim, 512)
- *How many more points did Jabu score than Jeremy in Round 1?* (144)
- *How many more points did Aareli score in Round 1 than in Round 2?* (59)
- *What is the total of Jeremy's and Jabu's scores for both rounds?* (921)

Unit introduction

Materials
Currency notes and coins for demonstration – you can draw and photocopy simplified versions and cut them out. You can use dollars and cents or local currency. If using local currency, remind the class that the course books use dollars and cents, and pounds and pence.

Teaching guidance
Begin by showing the children currency notes and coins and discussing these. They may remember the $1, $5, $10 and $20 notes from their work at Level 2.

If necessary, introduce the $50, $100 or larger denomination notes. Ask questions such as:
- *How much is this note worth? Which note is worth more than this? Which note is worth the most?*
- *How many different kinds of coins can you see?*
- *Which is worth the most money?*
- *Which is worth the least money?*
- *What can you buy with this much money? Do you think you can buy a pair of shoes with this much? What about a loaf of bread? What about a car? Why not?*
- *Would you rather have one of these or two of these? (You can vary this question using different notes and coins, for example: Would you rather have three 50c coins or two $1 coins? Why?)*

Explain the equivalence between 100c and $1 (or adapt this for your *local currency*). Ask questions about the coins and notes:
- *How many of these (showing a given denomination) do I need to make $1 (or $2 or $5)?*
- *How many of these (holding up a given coin) do I need to make the same amount as this (holding up a note)?*

Ask the children to note down the *cost* of three items that cost between, for example, $1 and $100 the next time they go shopping with their families. Check how the children recorded the amounts and make a class display of the different ways in which *money* amounts are written in everyday life. If this is not possible, find advertisements of prices and work from pictures.

Write money amounts

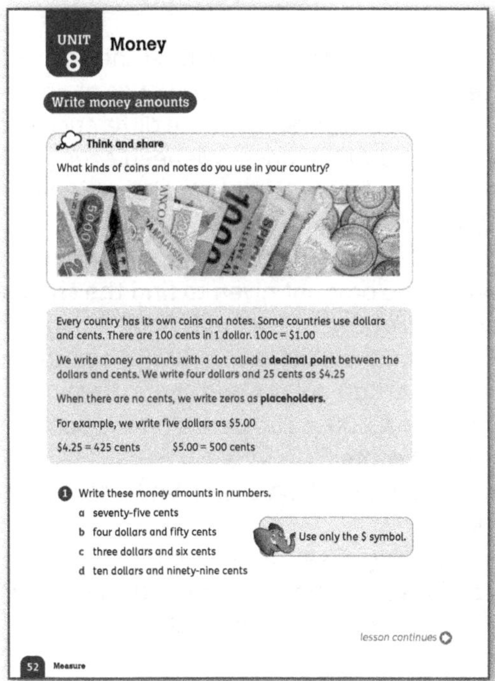

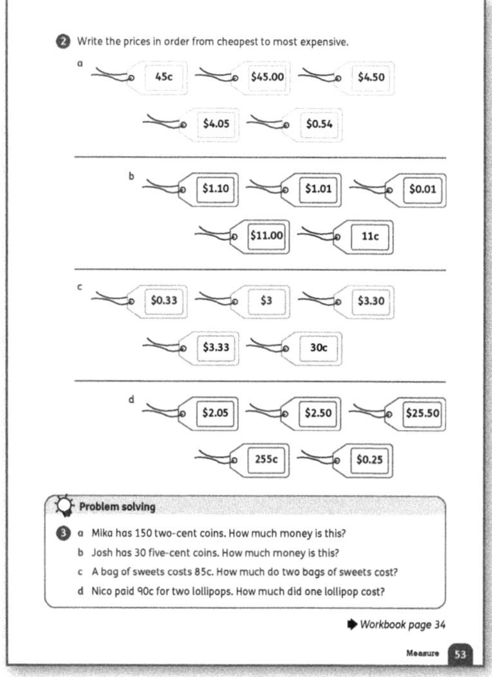

Materials
Local currency notes and coins (You can draw and photocopy simplified versions and cut them out.)

Warm-up
- <u>Think and share:</u> Look at the photos of coins and notes from different currencies on **Pupil Book 3 page 52** and discuss the questions on this page and on **page 53**. Remind the children of the different coins used locally.

Focus
- Work through the examples of money notation with the children. The children will be familiar with seeing prices written in figures from their everyday experience.
- Note that they do not formally have to learn the term *decimal point* at this stage, but it is essential to stress the importance of placing the dot (decimal point) in the correct position.
- Also show that we put a zero to show when there is nothing in a given place position.
- Teach the term *placeholder*. *We call zero the 'placeholder' because it holds the place for that position.* Demonstrate this for amounts with fewer than ten cents, such as $45.05 or $195.09. (You might like to ask: *What would happen if we didn't use zero as a placeholder?* Let the children give some examples.)
- Let the children work in pairs to practise writing sums of money. One child says an amount of money in words and the other writes it down in figures.
- Let the children work on their own to complete questions 1 and 2 on **Pupil Book 3 page 52**.
- <u>Problem solving:</u> The children can work in pairs to solve the word problems in question 3.

Follow-up
Let the children complete **Workbook 3 page 34** on their own and then have them work in pairs to compare their answers.

Challenge

Children who are comfortable with writing money amounts can do word problems in which they find totals of amounts. They should explore strategies of how to do this in a variety of ways – making amounts in *dollars* and *cents* by regrouping amounts of more than 100c as whole dollars. Some may adapt the column addition they learnt in the previous unit.

Support

Give additional examples of writing money amounts, focusing on practising the use of zero as a placeholder, for example:

- 9 dollars and 5 cents
- 9 dollars and 50 cents
- 50 dollars and 9 cents.

Interesting mistakes

The children have not yet learnt about decimals, so the use of the decimal point in currency notation may prove a little tricky for some children. Because they have not yet learnt about decimal fractions, you need to show that the numbers after the point are the cents (or local equivalent), and there can only be the two places. If we add more than 100 cents, it gets carried to the dollar column. Use the notation of hundreds, tens and ones rather than the idea of decimal tenths and hundredths.

Answers for Pupil Book 3 pages 52–53

Think and share: Individual answers

1. a 75c or $0.75   b $4.50
   c $3.06   d $10.99
2. a 45c, $0.54, $4.05, $4.50, $45.00
   b $0.01, 11c, $1.01, $1.10, $11.00
   c 30c, $0.33, $3, $3.30, $3.33
   d $0.25, $2.05, $2.50, 255c, $25.50
3. a $3   b $1.50
   c $1.70   d 45c

Answers for Workbook 3 page 34

1. Individual answers that add to the totals given.
   For example:
   a four $1 notes and one 50c coin
   b one $5 note, three $1 notes and one 50c coin
   c one $10 note, one $5 note, one $1 note, one 50c coin and one 10c coin
   d one $10 note, two $1 notes, one 50c coin, one 25c coin and one 5c coin
   e one $10 note, three $1 notes, one 50c coin and two 10c coins
2. a $5.50   b $1.50
3. a $4.40   b $7.20   c $6.30

Totals and change

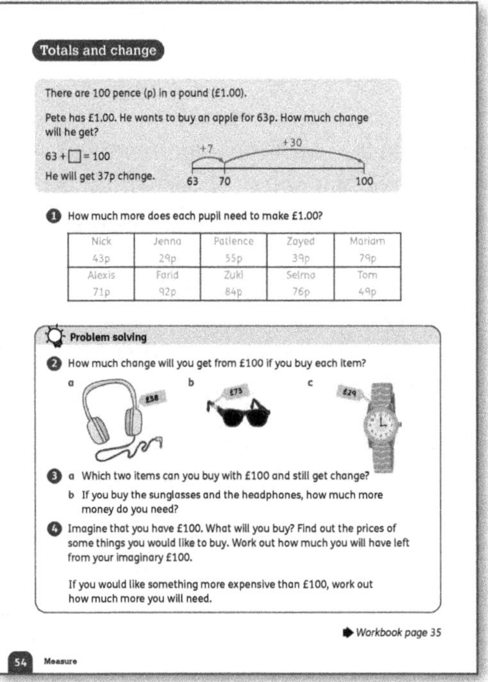

Materials

Photocopied or drawn and cut-out copies of UK coins (1p, 2p, 5p, 10p, 20p, 50p, £1)

Warm-up

Use items in the classroom and cut-out coins to model buying an item, paying for it and receiving change. For example: *This pencil costs 30 pence. I pay with a 50 pence coin. How much change do I get? (20 pence)*

Focus

- Make sure the children understand the vocabulary we use to talk about money. The *price* of an item is how much it costs. *Change* is the difference between what we pay and the price. We need change because sometimes we have a coin or note that is worth more than the price.
- Teach the children how to count on from the cost of an item to the amount given to find the change, as in the example at the top of **Pupil Book 3 page 54**.
- In this lesson, the children work with pounds and pence rather than dollars and cents. Remind the children that £1.00 = 100 pence (in the same way as $1.00 = 100 cents).
- Let the children work on their own to complete question 1.
- Problem solving: Give the children time to work on questions 2–4. Then ask them to share their strategies for solving these problems.

Follow-up

Use **Workbook 3 page 35** as a challenge to see how the children make each amount up to £1.00, double amounts and work out change. They can use photocopied or mocked-up copies of the UK coins to help them. Let them compare and discuss their answers.

Answers for Pupil Book 3 page 54

1 Nick: 57p   Jenna: 71p
Patience: 45p   Zayed: 61p
Mariam: 21p   Alexis: 29p
Farid: 8p   Zuki: 16p
Selma: 24p   Josh: 51p

2 a £62   b £27   c £71

3 a headphones and watch (The change would be £33.)
b £11

4 Individual answers

Answers for Workbook 3 page 35

1

Amount I have	Coins I need to make £1.00
25p	50p, 20p, 5p
75p	20p, 5p
52p	20p, 20p, 5p, 2p, 1p
88p	10p, 2p
90p	10p
18p	50p, 20p, 10p, 2p
39p	50p, 10p, 1p

2 a 90p   b £1.80   c £1.60
d £1.70   e £3   f £7

3 a 75p   b 55p   c £1.50
d £1.90   e £1.01   f 27p

End-of-unit check

Use these activities to check the work on money covered in this unit.

- Give the children collections of coins and notes and have them work out the total amount.
- Give amounts in words for the children to write in numbers, or say an amount and have them write it using the correct money notation.
- Ask the children to write a given price using money notation.
- Ask questions such as:
  - *How many more cents do I need to make $1 if I have (give an amount)?*
  - *I buy a banana that costs 27p. How much change do I get from £1? (73p)*
  - *I paid for an item with $1 and got 35c change. How much did the item cost? (65c)*

UNIT 9   Mass

Learning objectives

- Estimate, measure, compare, add and subtract mass in grams (g) and kilograms (kg).
- Use instruments that measure mass.
- Understand the relationship between units.

Key words

mass   weight   kilogram   gram   balance
scale   weigh   heavy/heavier/heaviest

Unit introduction

Materials

Plastic bag, a 1-kg mass (such as a bag of sugar or flour or a 1-litre plastic bottle filled with water); a variety of objects to heft and weigh, including: 1-kg and 2-kg bags of flour or rice; objects with a range of masses of more or less than 1 kg; materials for making a simple balance: coat hanger, two bags, bulldog clips; balance scales; 1-kg, $\frac{1}{2}$-kg and $\frac{1}{4}$-kg weights; spinner marked 'weighs more' and 'weighs less'; kitchen scale calibrated in kilograms

The correct term for a measurement in kilograms and grams is *mass*, but people often use the word *weight* in everyday language.

Teaching guidance

At this stage, the children only work with *kilograms* and *grams* and they need to understand that there are 1000 grams in a kilogram. They should also realise that 500 g is equivalent to half a kilogram. The children do not need to multiply and divide by 1000 at this stage, so they will not work with conversions between units beyond expressing mass in different ways.

Some children may not be familiar with measuring scales and may still struggle to read them or to work out what the intervals are. You may need to spend some time teaching them how to do this before you ask them to read and record measurements.

By now, the children are used to word problems, but they may forget to include the units (or even to consider the units) when they have to solve problems involving measures. Discuss problems with the class and talk about what they are being asked to do, to help them see the importance of including units (kg or g) in their answers.

Here are some ideas for introductory activities for this unit.

- Set up a 'hefting' activity. Hefting means holding objects to compare or guess their mass. Give the children a plastic bag with a 1-kg mass in it (see the suggestions in the Materials section, above). Let them hold this in one hand and pick up some items in the other hand. They should compare the mass and say whether each item hefted is 'less than', 'about equal to' or 'more than' 1 kilogram.
- The children can make a simple *balance* by using bulldog clips to attach a bag to each end of a coat hanger, then suspending the coat hanger from a hook or nail. A coat hanger balance must be suspended away from the wall so it is free-moving. Give the children a group of objects. The children can estimate the mass of each object and then use 1-kg, $\frac{1}{2}$-kg and $\frac{1}{4}$-kg *weights* to check the accuracy of their estimates. They can place the objects in order of increasing mass.
- The children can work in pairs with a 'weighs more'/'weighs less' spinner. One child chooses an object and spins; the second child has to identify an object that weighs more or weighs less than the other child's object. They can check using a balance.
- The children can use a kitchen scale that is calibrated in kilograms to measure the mass of some objects. Spend time talking about how to read the scale and what the divisions mean.

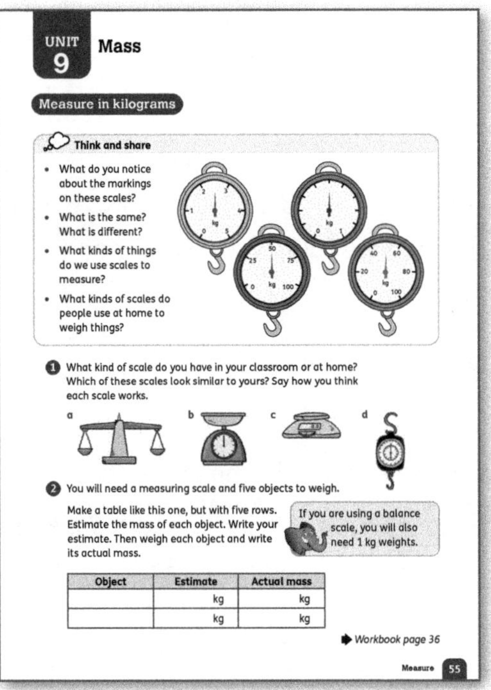

Materials

Different kinds of scales or pictures of different scales; kitchen scale or other *scale* that can measure in kilograms; objects that the children can *weigh* in kilograms (for example, a bag of potatoes, bag of apples, bag of rice, large toy, large book, box of toys)

Warm-up

- <u>Think and share:</u> Discuss the picture of different scales in this section of **Pupil Book 3 page 55**. Ask the children to describe what they can see – what is similar and what is different. You can ask questions such as:
  - *What can we measure with this kind of scale?*
  - *What do the markings on the scales show us?*
  - *What is the heaviest mass this scale can measure? (What mass does it go up to?)*
  - *Which scale can measure a heavier object than that?*
  - *What is the heaviest object this scale can show?*
  - *Where is the arrow pointing?*
  - *What measurement does the arrow show?*
  - *Where would the arrow point if the mass was (1 kg/2 kg/20 kg, etc.)?*
  - *What mass would move the arrow from here to here (pointing to the calibrations)?*
- The children should notice that the scales are calibrated in different ways – one can measure up to 5 kg, one can measure up to 1 kg, and two can measure up to 100 kg but they are marked in different ways.
- You can also practise skip counting through the calibrations on these scales.

Focus

- In question 1 on **Pupil Book 3 page 55**, the children discuss the different kinds of scales shown and where they might use them. The first is a balance scale, which works using weights and balances. When the pans are level, the masses are equal. The next two are kitchen scales – one works by the mass pushing down on a lever that is connected to the needle on the scale; the other is electronic and measures using sensors. The hanging scale works by the mass pulling down on the hook – the mass is shown on the dial.
- In question 2, the children estimate the masses of a range of objects and then measure to check. They copy and complete the table.

Follow-up

The children can use **Workbook 3 page 36** to carry out and record further practical work with mass.

Support

- Some children may need to revise counting and skip counting to 1000.
- Revise the relationship between grams and kilograms.
- Show items of approximately 1 kilogram and let the children feel how much they weigh. If you have balance scales in the classroom, let the children use objects to make up masses of 1 kg.

Answers for Pupil Book 3 page 55

<u>Think and share:</u> Possible answers:

The markings are different on the scales – one scale can measure masses up to 5 kg, one up to 1 kg, and two can measure up to 100 kg, but they are marked in different ways.

The same: The two scales at the bottom measure up to 100 kg. They all measure by hanging objects from a hook at the bottom. They all have a dial with a pointer that shows the mass.

Different: One scale can measure masses up to 5 kg, one up to 1 kg, and two can measure up to 100 kg.

Possible answers for uses of scales at home: kitchen scales to weigh foods; bathroom scales to weigh people; luggage scales to weight suitcases.

1 Possible answers:

  a Place an object in one pan. In the other pan, place weights to make the pans balance. The mass of the weights tells you the mass of the object.

  b Place something in the pan. The pointer on the dial points to the mass.

  c Place something on the surface of the scale. The display shows the mass.

  d Hang something on the hook. The pointer on the dial points to the mass.

2 Individual answers

Kilograms and grams

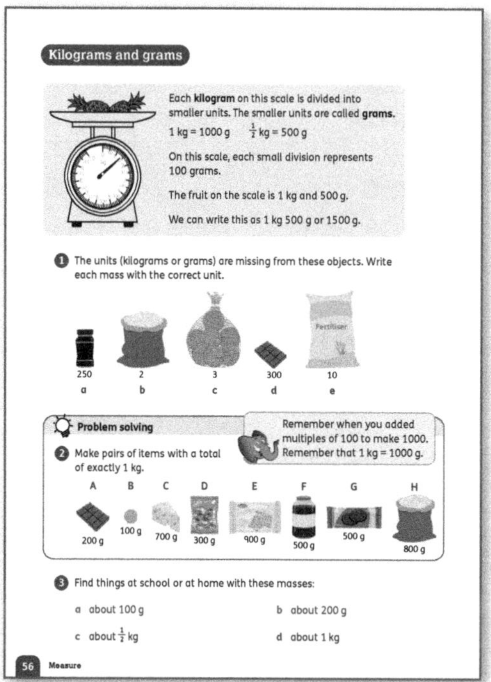

Materials

Scales and objects for weighing

Warm-up

- Start with practical work. Let the children measure the mass of different objects to make up masses of 1 kg and half a kilogram (500 g).
- While they do this, discuss objects that we would measure in grams and objects that we would measure in kilograms.
- Ask: *Would you measure this in grams or kilograms?*
- Name some objects and let the children decide which unit they would use, for example: a ruler (g), a watermelon (kg), a phone (g), an earring (g), a sock (g), a suitcase full of clothes (kg).

Focus

- Once you have done some practical work with scales and the divisions on them, work though the example at the top of **Pupil Book 3 page 56** with the class.
- Then let the children work independently to complete questions 1–3.
- <u>Problem solving:</u> In question 2, the children find pairs of items with a total mass of 1 kilogram.
- Question 3 is a practical activity that could be done in the classroom or set as homework.

Read scales

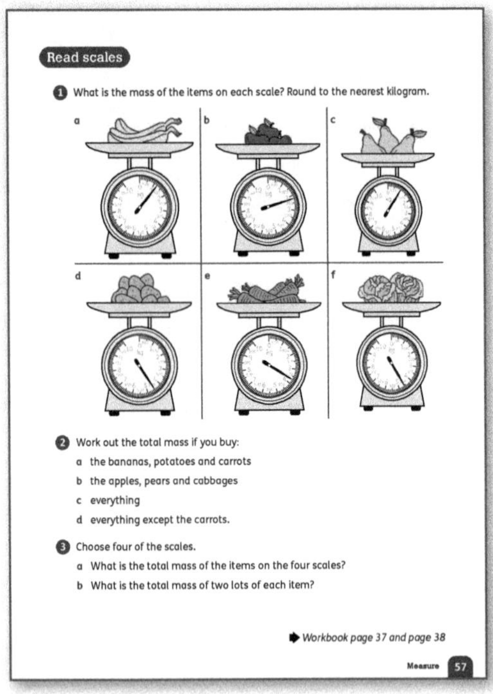

Materials

Scales and objects for weighing

Warm-up

Spend time making sure the children understand how to read mass on a calibrated scale before they complete this page. You can do this practically, using a scale and a range of objects.

Focus

- The children can work individually or in pairs on the questions on **Pupil Book 3 page 57**.
- In question 1, the children read the mass shown by the needle on each scale. Note that all the scales have the same calibrations and all go up to 10 kg.
- The children need to add up the rounded amounts (and subtract for part d) in question 2.
- For question 3, the children pick any four of the scales from question 1 and add up the total mass of the items. For question 3b, they have to work out double the mass on their chosen four scales.

Follow-up

- Use **Workbook 3 page 37** and **page 38** to consolidate and assess work on reading scales.
- For **Workbook 3 page 38** question 2, ask some questions about the table to check understanding before the children work on this question in pairs.

- You may also need to explain the meaning of the word *maximum* (the greatest amount that is allowed or possible).

Challenge

Ask the children to solve this problem:
- *Juanita needs to put 14 boxes in her van.*
- *The boxes have the following masses: 5 kg, 10 kg, 12 kg, 14 kg, 15 kg, 16 kg, 17 kg, 18 kg, 19 kg, 20 kg, 21 kg, 23 kg, 24 kg and 26 kg.*
- *The most she can take in one trip is 40 kg.*
- *On her first trip, she takes the 20 kg box, the 15 kg box and the 5 kg box.*
- *She needs to make 5 more trips.*
- *Write the masses of the boxes that she takes each trip.* (In any order: 14 kg and 26 kg; 24 kg and 16 kg; 23 kg and 17 kg; 21 kg and 19 kg; 18 kg, 12 kg and 10 kg)

Support

Some children may need help to adapt their rounding skills when working with grams and kilograms. Explain how to do this: *If there are 500 g or more, round up to the next kilogram. Otherwise round down.* Demonstrate this using a number line.

To assess the work covered on mass in this unit, ask questions such as:
- *What does kg stand for?* (kilograms)
- *How many grams are there in a kilogram?* (1000)
- *How many grams are there in a half kilogram?* (500)
- *How many grams are there in a quarter kilogram?* (250)
- *How many 500-g packets of butter have the same mass as a 1-kg packet?* (2)
- *How many 200-g packets of pasta have the same mass as a 1-kg packet?* (5)

- *If a 1-kg block of cheese is cut into four equal pieces, what is the mass of each piece?* (250 g)
- (Give a list of masses in grams or kilograms.) *Arrange these masses in order from smallest to greatest.*

Give the children items to weigh and ask them to tell you the mass to the nearest kilogram.

Give the children two objects and ask: *Which of these two objects is heavier/lighter?* (The children can hold the objects to decide.)

Ask: *Which is heavier, 1 gram or 1 kilogram?* (1 kilogram)

 UNIT 10 Multiplication and division

Learning objectives
- Understand and explain the relationship between multiplication and division.
- Understand and explain commutative and distributive properties of multiplication and use these to simplify calculations.
- Know 1, 2, 3, 4, 5, 6, 8, 9 and 10 times tables and related division facts.
- Estimate and multiply/divide by 2, 3, 4 and 5 for whole-number products up to 100.
- Use knowledge of factors and multiples to understand tests of divisibility by 2, 5, 10, 25, 50 and 100.
- Multiply numbers by 10 using knowledge of place value.
- Multiply 2-digit numbers by 1-digit numbers, using mental and formal written methods.
- Solve problems, including missing number problems, involving multiplication and division, including positive integer scaling problems and correspondence problems in which *n* objects are connected to *m* objects.
- Count from 0 in multiples of 4, 8, 50 and 100.
- Apply known multiplication and division facts to solve contextual problems with different structures, including quotative and partitive division.

Key words
array   times table   repeated addition   times
multiplication   multiply   multiple   division
divide   group   share   share equally
remainder   inverse operation   product
factor   divisible   divisibility

Materials
Counters, beans or other small objects; pictures of groups of objects; number lines and number tracks

Teaching guidance

> Do not introduce division and multiplication at the same time. Develop the concept of multiplication as repeated addition to build up multiples. Once the children have firmly grasped this concept, move on to division – breaking whole multiples into smaller groups.

Here are some ideas for practical activities in multiplication and division:
- The children can use small objects (beans, counters, etc.) to make *arrays*:
  - The children arrange the objects in pairs. They add up the number of objects in 1 pair, 2 pairs, etc.
  - The children arrange the objects in sets of 3, 4 and 5. They add up the number of objects in 1 set, 2 sets, etc.
- Show the children pictures of grouped objects (for example, a pair of socks, a 3-wheeled tricycle, an ox with 4 legs and a pack of 5 ice lollies). Let them say how many socks, wheels, legs, ice lollies, etc. there would be if each child in their group had one of the objects shown. Let them explain how they worked this out.
- Put the children in groups of five. Hand out the same number of counters to each child. Ask them to skip count to find out how many counters there

are altogether in their group. They can record their results in a table like this one (which uses the example of 3 counters per child):

Number of children	1	2	3	4	5
Number of counters	3				

Vary the number of counters to develop the concept of *repeated addition* and multiplication for the *times tables* you are teaching.

- Ask the children to name things that come in threes, such as wheels on a tricycle. Get the children to make 10 groups of 3 using counters. Ask the class how many counters there are in 1 group of 3, 2 groups of 3, etc. Write the total on the board each time.
- Ask the children to explain the pattern. Reinforce the pattern by counting in threes on a number line. Show the children that, starting at 0 and jumping 3 each time, they get the same pattern of numbers as when they added the sets of 3 together.
- Summarise the 3 times table in a whole-class session. Show the children that 2 groups of 3 can be written as 2 × 3, etc. Identify × as the sign we use for multiplication.
- Repeat this activity for other times tables.

Revise multiplication and division

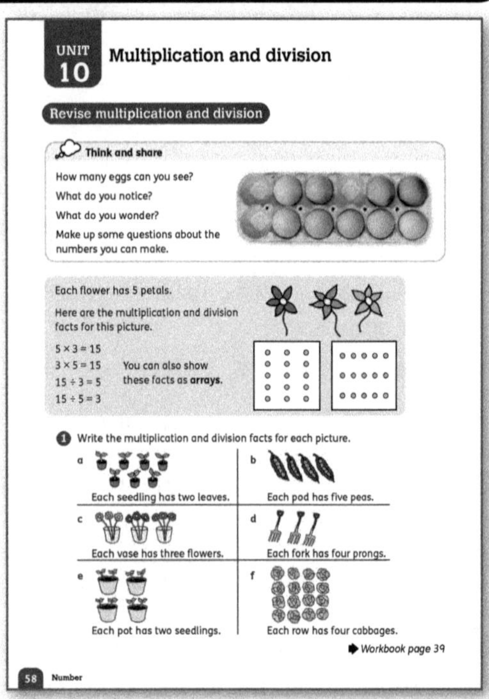

Materials

Counters, small objects or Numicon number frames (pages 21–22) for creating arrays; pictures of arrays; cards or objects in groups of the same size to demonstrate repeated addition

The process of repeated addition generates multiples of the starting number. For example:

Start at 0. Add 4. Repeat the addition 5 times.
$0 + 4 + 4 + 4 + 4 + 4 = 20$

We can show that the repeated additions generate multiples of 4:

	0	
First multiple	+ 4	= 4
Second multiple	+ 4	= 8
Third multiple	+ 4	= 12
Fourth multiple	+ 4	= 16
Fifth multiple	+ 4	= 20

We added 4 five times.
5 *times* 4 = 20
We can also show this as an array:

The array shows that 5 rows with 4 dots in each row makes 20.

It also shows that 4 columns with 5 dots in each column makes 20.

It is very important that the children have a strong understanding that *multiplication* means building up a number by repeated addition or working through *multiples*. It is also worth demonstrating with pictures or objects that multiplying usually means making more, by doubling, tripling or 'timesing' by any other number.

The children must have this firm concept of multiplication before they work with the inverse, *division*. As you introduce (or revise) the division sign, emphasise that *multiplying* builds up by repeating an amount or quantity. Division breaks down by splitting up an amount or quantity into smaller, equal bits.

Warm-up

<u>Think and share:</u> Talk about the picture of the egg box on **Pupil Book 3 page 58** with the children. Ask what they see and notice and what they wonder about the picture. You can ask a wide range of questions about these kinds of pictures to introduce the children to different ways of seeing numbers, for example:

- *How many eggs can you see?* (9)
- *What shapes or arrangements do you notice?*
- *What shape does it make to you? Does anyone see it differently?*
- *How many eggs are missing?* (3)

- *How would you show this as a number sentence?*
- *How could you share the eggs into groups? Could you make two equal groups? Why not?* (9 is not an even number)
- *Could you make three groups?* (Yes, 3 groups of 3)
- *If the egg box was full, could you split the eggs into groups the same way?* (Yes, you could split the eggs into 4 groups of 3.)
- *What other ways could you share the eggs from a full egg box?* (12 groups of 1, 2 groups of 6, 3 groups of 4, 6 groups of 2 and 1 group of 12)

Focus
- Move on to the picture of the flowers and the arrays. Unlike the egg box picture, arrays are shown in neat rows and columns of equal size. Ask: *What is the same about the two arrays?* (Both show 15 dots) *What is different?* (One shows 3 columns of 5 and the other shows 3 rows of 5)
- Demonstrate how we repeat add or build up to multiply. Division is the opposite process – when we *divide*, we take away *groups*, or *share* into equal groups.
- For question 1, work through the first one or two questions on multiplication and division facts with the class. Then let the children continue independently if possible.
- For question 1d, you may need to explain that the individual points of a fork are called *prongs*.

Follow-up
Use **Workbook 3 page 39** to reinforce the work on multiplication and division facts.

Challenge
Some children may want to work out multiplication and division facts for greater numbers. You can give the children arrays of 48 counters, 64 counters, 72 counters and 81 counters, and let the children find as many different divisions and multiplications as they can.

Another good challenge activity is to link multiplication to patterns and sequences.
- Start with a sequence of multiples, for example: 2, 4, 6, 8, …
- Ask: *In this sequence, we add 2 to get the next term. If we multiply by 2 to get the next term instead of adding 2, what pattern do you think it will make? Do you think any of the terms will be the same?* (No) *Do you think the sequence will grow in the same way?* (No) *Do you expect that we will get odd or even numbers?* (Even)
- Let the children experiment with continuing the pattern by multiplying by 2. They will generate the sequence 2, 4, 8, 16, …
- Let them find the first ten terms. Ask: *What do you notice? What is the same? What is different? Why do the two patterns change so much?*
- Let the children try the same with threes (start with 3 and add 3 to get the next term; then start with 3 and multiply by 3 to get the next term; this will generate the sequence 3, 9, 27, …).

Support
Use counters or number frames to set out arrays and help the children to see the commutative property of multiplication (3×4 has the same product as 4×3).

Interesting mistakes
Some children may get confused about the difference between dividing and multiplying and write calculations such as '$7 \div 2 = 14$'. This can be very difficult to explain. Use arrays and emphasise the word 'times' for multiplying:
- *How many times do we add another 7? 2 times. 2 times 7 is 14. Let's look at the 2s. How many times do we add another 2? 7 times 2 is also 14.*
- *When we divide, we break up a number into equal bits. Can we divide 7 into groups of 2? How many groups would it make?* (3 and 1 left over) *Can we divide 14 into groups of 2?*

Continue with similar questions.

Answers for Pupil Book 3 page 58

<u>Think and share:</u> 9 eggs
Individual answers. For example:
I notice that there are some eggs missing.
I wonder how many eggs were in the box when it was full.
Example questions: How many eggs in a full box? ($2 \times 6 = 12$) How many groups of 3 eggs are there? ($9 \div 3 = 3$) How many groups of 3 eggs are there in a full box? ($12 \div 3 = 4$)

1 a $7 \times 2 = 14$, $2 \times 7 = 14$, $14 \div 7 = 2$, $14 \div 2 = 7$
   b $4 \times 5 = 20$, $5 \times 4 = 20$, $20 \div 4 = 5$, $20 \div 5 = 4$
   c $3 \times 3 = 9$, $9 \div 3 = 3$
   d $3 \times 4 = 12$, $4 \times 3 = 12$, $12 \div 3 = 4$, $12 \div 4 = 3$
   e $4 \times 2 = 8$, $2 \times 4 = 8$, $8 \div 4 = 2$, $8 \div 2 = 4$
   f $4 \times 4 = 16$, $16 \div 4 = 4$

Answers for Workbook 3 page 39

1 a $5 \times 2 = 10$, $10 \div 2 = 5$, $2 \times 5 = 10$, $10 \div 5 = 2$
   b $2 \times 4 = 8$, $8 \div 2 = 4$, $4 \times 2 = 8$, $8 \div 4 = 2$
   c $4 \times 3 = 12$, $12 \div 3 = 4$, $3 \times 4 = 12$, $12 \div 4 = 3$
   d $4 \times 5 = 20$, $20 \div 4 = 5$, $5 \times 4 = 20$, $20 \div 5 = 4$
   e $10 \times 2 = 20$, $20 \div 2 = 10$, $2 \times 10 = 20$, $20 \div 10 = 2$
   f $8 \times 2 = 16$, $16 \div 8 = 2$, $2 \times 8 = 16$, $16 \div 2 = 8$

Times tables

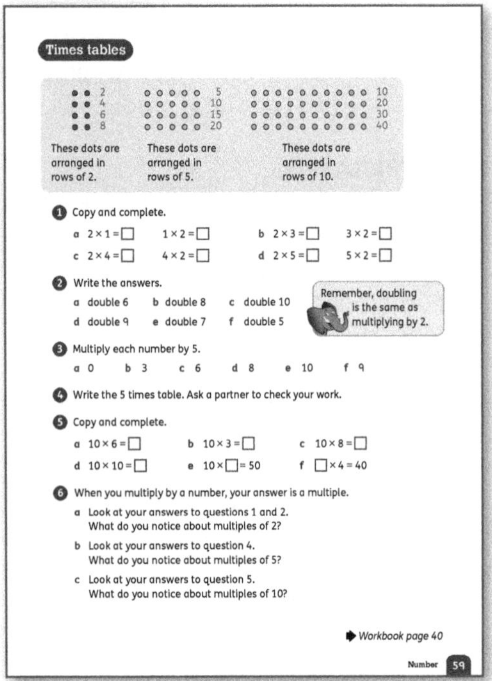

Materials

Counters or Numicon number frames (pages 21–22) for creating arrays; clean, empty yoghurt pots

Warm-up

- Look at the examples at the top of **Pupil Book 3 page 59** with the class. Ask the children to describe what they see. Ask guiding questions such as:
  - *How are the red dots arranged? How many are there in the first row?* (2)
  - *How many do the dots in the first and second rows make together?* (4)
  - *How can we say that as a plus sum?* (2 + 2 = 4)
  - *How many 2s are we adding together to make 4?* (Two. Two 2s make 4.)
- Continue to discuss the array in this way to make 6 and 8. Then discuss the arrays with rows of 5 and rows of 10 in the same way.

Focus

- Spend some time talking about multiples of 0.
- Write the number 0 and the word *zero* on the board and ask: *What do I get if I start with 0, and add 0?* (0) *What if I add another 0?* (0) *So, 2 times 0 is 0. 3 times 0 is also 0. Let's add a fourth 0. What are 4 zeroes?*
- Ask: *Is there any number we can multiply by 0 to get a different answer?*
- Ask the children if they can find other ways to explain or show this. Guide them to notice that any number multiplied by 0 is equal to 0.
- Let the children work independently through questions 1–6 on **Pupil Book 3 page 59**.

Follow-up

The children can complete **Workbook 3 page 40** independently for further practice of the the 2, 5 and 10 times tables, or you can use this page to assess the work covered in this lesson.

Challenge

The children can explore other times tables and work on finding patterns that repeat (for example, in the 8 times and 9 times tables).

Support

If children need a different concrete model, use yoghurt pots with the same number of counters or small objects in each. Say, for example:
I have 1 pot with 5 counters. The total number of counters is 5. Let's add one more pot. 1 five, plus another five … I have 2 fives.
5 + 5 = 10
2 fives = 10
2 pots times 5 counters equals …? (10)

As you demonstrate using concrete materials, write the times table on the board or in a chart.

Answers for Pupil Book 3 page 59

1 a 2, 2   b 6, 6
    c 8, 8   d 10, 10

2 a 12   b 16   c 20
    d 18   e 14   f 10

3 a 0   b 15   c 30
    d 40   e 50   f 45

4
$1 \times 5 = 5$ $7 \times 5 = 35$
$2 \times 5 = 10$ $8 \times 5 = 40$
$3 \times 5 = 15$ $9 \times 5 = 45$
$4 \times 5 = 20$ $10 \times 5 = 50$
$5 \times 5 = 25$ $11 \times 5 = 55$
$6 \times 5 = 30$ $12 \times 5 = 60$

5 a 60   b 30   c 80
    d 100   e 5   f 10

6 a Multiples of 2 are always even numbers. They end in 0, 2, 4, 6, or 8.
    b Multiples of 5 always end in either 0 or 5.
    c Multiples of 10 always end in 0. Every second multiple in the 5 times table is a multiple of 10.

Answers for Workbook 3 page 40

1

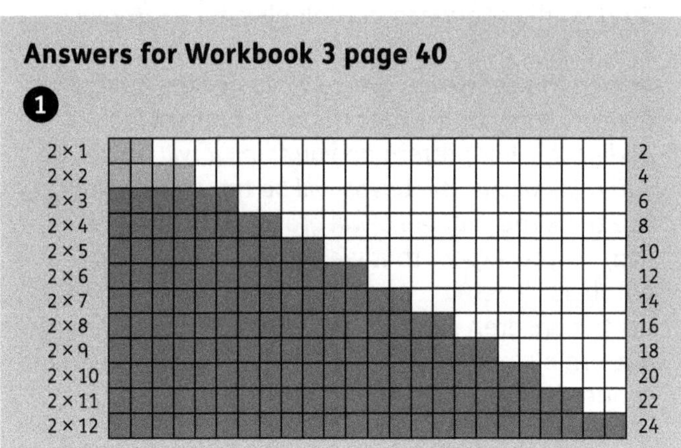

2 a–c

1	2	3	4	(5)	6	7	8	9	(10)
11	12	13	14	(15)	16	17	18	19	(20)
21	22	23	24	(25)	26	27	28	29	(30)
31	32	33	34	(35)	36	37	38	39	(40)
41	42	43	44	(45)	46	47	48	49	(50)
51	52	53	54	(55)	56	57	58	59	(60)
61	62	63	64	(65)	66	67	68	69	(70)
71	72	73	74	(75)	76	77	78	79	(80)
81	82	83	84	(85)	86	87	88	89	(90)
91	92	93	94	(95)	96	97	98	99	(100)

d Individual answers. For example: all the shading, circles and underlining are in columns. All the numbers ending in 0 are shaded, circled and underlined.

More times tables

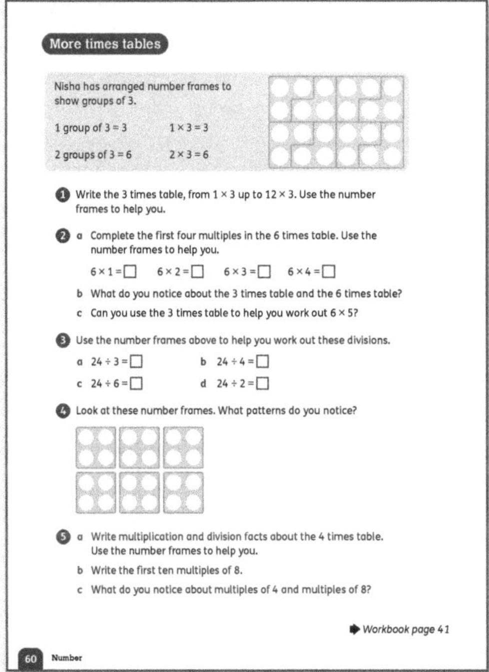

Materials

Numicon number frames (pages 21–22) or counters to continue building up times tables visually

Warm-up

- Turn to **Pupil Book 3 page 60**. Discuss the picture of the number frames with the class. Ask questions such as:
  - *How many holes are there in each frame?*
  - *How are the frames arranged?*
  - *What do you notice?*
  - *How can you count up the holes?*
  - *What do you wonder?*
- The children may make observations such as: 'Each frame is a set of 3, arranged in an L-shape. Each rectangle is made of 2 sets of 3. So, we can see the picture as 8 sets of 3 or as 4 sets of 6.'

Focus

- Ask the children to use number frames or counters to make groups of 9, and work out the first 12 multiples of 9. They can write these as the 9 times table.
- Note that the 7, 11 and 12 times tables will be introduced at Level 4, but you might like to start introducing the children to these tables at Level 3.
- Work with the children through questions 1–5, or let them complete these independently.

Follow-up

Use **Workbook 3 page 41** to consolidate the 3 times table.

Answers for Pupil Book 3 page 60

1
$1 \times 3 = 3$	$7 \times 3 = 21$
$2 \times 3 = 6$	$8 \times 3 = 24$
$3 \times 3 = 9$	$9 \times 3 = 27$
$4 \times 3 = 12$	$10 \times 3 = 30$
$5 \times 3 = 15$	$11 \times 3 = 33$
$6 \times 3 = 18$	$12 \times 3 = 36$

2 a 6, 12, 18, 24

   b Possible answers: Every other number in the 3 times table is in the 6 times table. Each multiple of 6 is double the same multiple of 3.

   c $6 \times 5 = 3 \times 10 = 30$

3 a 8   b 4   c 6   d 12

4 Possible answers: There are 6 frames of 4. There are 4 rows of 6. There are 3 columns of 8.

5 a Possible answers: $6 \times 4 = 24$, $4 \times 6 = 24$, $24 \div 4 = 6$, $24 \div 6 = 4$

   b 8, 16, 24, 32, 40, 48, 56, 64, 72, 80

   c Possible answers: Every other number in the 4 times table is in the 8 times table. Each multiple of 8 is double the same multiple of 4.

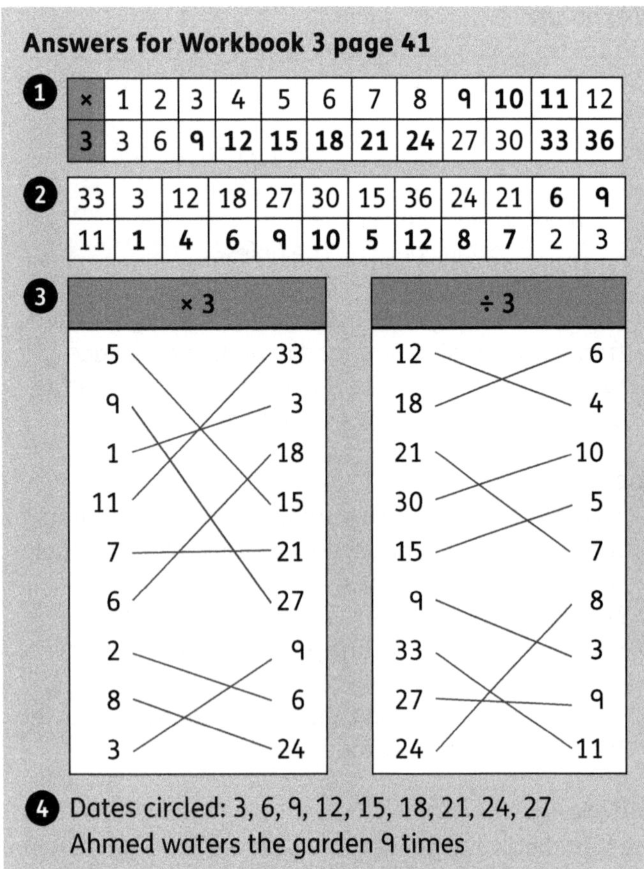

Answers for Workbook 3 page 41

1

×	1	2	3	4	5	6	7	8	9	10	11	12
3	3	6	9	12	15	18	21	24	27	30	33	36

2

33	3	12	18	27	30	15	36	24	21	6	9
11	1	4	6	9	10	5	12	8	7	2	3

3

× 3	÷ 3

(matching activity: × 3 — 5, 9, 1, 11, 7, 6, 2, 8, 3 matched to 33, 3, 18, 15, 21, 27, 9, 6, 24; ÷ 3 — 12, 18, 21, 30, 15, 9, 33, 27, 24 matched to 6, 4, 10, 5, 7, 8, 3, 9, 11)

4 Dates circled: 3, 6, 9, 12, 15, 18, 21, 24, 27
Ahmed waters the garden 9 times

Revise division into groups

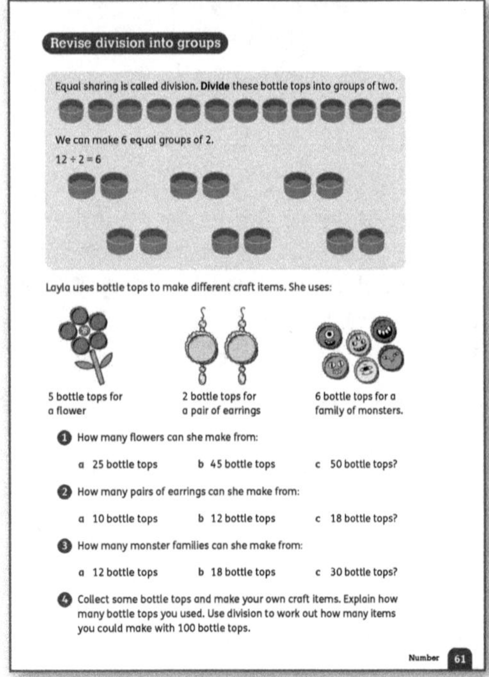

The children need to understand the concept of repeated subtraction in order to understand division into equal groups. You may need to show the children how to take away repeated groups using physical objects before, or alongside, work on a number line.

As you progress, the children will also work with repeated subtraction that leaves a *remainder*. This should be a fairly natural progression, as the fact that you cannot take another group of the same size away shows clearly that there are some left over.

Begin with practical work dividing objects into equal groups. The children could explore different ways of dividing a number into equal-sized groups. The numbers 12, 16, 24 and 36 will provide the children with many ways. Observe the children to see who still uses one-by-one sharing and who has moved onto grouping.

The children could use small objects (such as beans, bottle tops or counters) or a number line to find the number of groups of a particular size in a number. Using small objects, they can simply arrange them into groups and count the number of groups. Using a number line, they start at the number and move back in jumps equal to the size of the group until they reach 0. The number of groups is equal to the number of jumps.

The children could divide different numbers of beans into groups of 2 and discover which numbers leave a remainder of 1 and which do not. From this, they could identify a pattern of even and odd numbers.

Materials
Small objects such as counters and beans; number lines; bottle tops and craft materials

Warm-up
* Ask the children to use beans (or counters) to discover how many different ways they can share 12 beans equally between 2, 3, 4 and 6 children. Write their answers on the board and discuss them:
  2 children get 6 beans   3 children get 4 beans
  4 children get 3 beans   6 children get 2 beans.
* Introduce the idea of dividing things into groups. Remind the children of the different ways they were able to share 12 beans equally and show them that:
  12 is 2 groups of 6   12 is 3 groups of 4
  12 is 4 groups of 3   12 is 6 groups of 2.
* They should use beans to prove to themselves that this is true. Use a number line to show how a number can be divided into groups of equal size. For example, start at 16 and move back in jumps of 4 to 0. We need 4 jumps, so we can make 4 groups.

Focus
* Rewrite the different ways of sharing 12 equally using the ÷ sign:
  $12 ÷ 2 = 6$ $12 ÷ 3 = 4$ $12 ÷ 4 = 3$ $12 ÷ 6 = 2$
* Use a simple example to demonstrate that it is sometimes not possible to *share equally* and some

items may remain at the end of the sharing. For example, if we share 13 biscuits between 4 children they will each get 3 biscuits but there will be 1 biscuit remaining.

- Introduce the term 'divided by' to mean the same as 'shared equally among'. Division sentences can be written with both, for example '16 shared equally among 4' can be written as '16 divided by 4'.
- Turn to **Pupil Book 3 page 61** and ask the children to work through questions 1–4 about Layla's crafts. Provide the children with bottle tops to make their own craft items.

Challenge

Expand the questions from this page into more complex problem solving, for example:

- *Layla makes a pair of earrings. For each pair, she uses 2 bottle tops, 3 silver beads and 4 coloured beads. Work out how many of each she needs to make 8 pairs of earrings.* (16 bottle tops, 24 silver beads and 32 coloured beads)
- *Layla has 10 bottle tops, 12 silver beads and 20 colour beads. How many pairs of earrings can she make?* (4)

Support

Give the children a collection of bottle tops or counters and ask them to arrange them in arrays (rows and columns) to help work out how we can arrange them in equal sets. Let them explore other arrays that they can make with the same total.

Answers for Pupil Book 3 page 61

1	a 5		b 9		c 10
2	a 5		b 6		c 9
3	a 2		b 3		c 5
4	Individual answers				

Multiplication and division facts

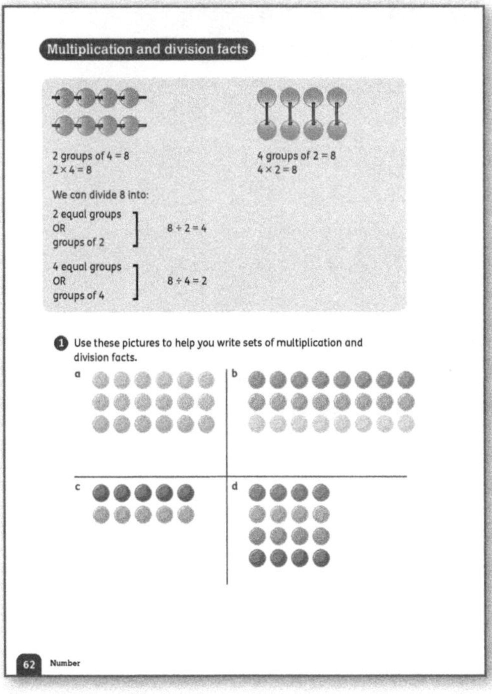

Materials

Small objects for creating arrays

The children are starting to solve division calculations using division facts derived from multiplication tables. They also need to be able to use these facts flexibly to solve different kinds of division:

- quotative (grouping) – for example: *There are 12 eggs in a box. I use 2 eggs for each omelette. How many omelettes can I make?* (12 ÷ 2 = 6)
- partitive (sharing) – for example: *There are 12 eggs in a box. I divide them equally between two bowls. How many eggs are in each bowl?* (12 ÷ 2 = 6)

Note that the children do not need to know the terminology for the different types of division, but your language should focus on mathematical relationships (*6 times 2 is 12, so 12 divided by 2 is 6*) as well as the different ways of making sense of the divisions (*12 divided into groups of 2 is equal to 6 groups. 12 divided into 2 equal groups makes 6 in each group*).

Warm-up

- Show the children a 3 by 5 array of objects. Divide the array into 3 rows of 5 and write 15 ÷ 3 = 5. Then divide the array into 5 columns of 3 and write 15 ÷ 3 = 5.
- Arrange the class into pairs and give each pair 18 small objects. Ask the pairs to make an array using all 18 objects, and then to write the division facts for their array. Then share answers and make sure the children have found the 2 by 9 and 6 by 3 arrays.

Focus

Go through the example on **Pupil Book 3 page 62**, then have the children work independently to write a set of multiplication and division facts for each representation in question 1.

Challenge

Give additional problem-solving questions, such as:

- *I have some peaches. I divide them equally between 2 plates, and there are 7 on each plate. How many peaches did I have to start with?* (14)
- *Nicky has a pencil case full of coloured pencils. She makes groups of 5 pencils. She has 6 groups. How many coloured pencils did she have to start with?* (30)
- *Tina builds a castle out of blocks. Each row in one wall has 6 blocks. A finished wall has 36 blocks. How many rows did Tina make to build the wall?* (6)

Support

- Some children may not have learnt all their multiplication and division facts by heart for instant recall.
- In the past, teachers would insist that without rapid recall of these facts, children would undoubtedly fall behind. However, we now know that

most people can recall some multiplication facts more easily than others.
- Typically, the 2 times, 5 times and 10 times tables are the easiest to recall. A child may easily recall that $3 \times 5 = 15$, but may be more hesitant about 3×6.
- Encourage the children to use a known fact to find another fact. For example, if we know that 3×5 is 15, we can add one more 3 to make 6 threes (3×6).

Interesting mistakes

The language of multiplication and division can lead to some interesting mistakes – we say 'divide by' and 'divide into' as well as 'times by' and 'multiply by'. Make sure you direct the children's attention to making sense of the problems they are solving rather than quickly trying to answer using known facts.

> ### Answers for Pupil Book 3 page 62
>
> 1 a $3 \times 6 = 18$, $6 \times 3 = 18$, $18 \div 6 = 3$, $18 \div 3 = 6$
>   b $3 \times 8 = 24$, $8 \times 3 = 24$, $24 \div 8 = 3$, $24 \div 3 = 8$
>   c $2 \times 5 = 10$, $5 \times 2 = 10$, $10 \div 5 = 2$, $10 \div 2 = 4$
>   d $4 \times 4 = 16$, $16 \div 4 = 4$

Division facts

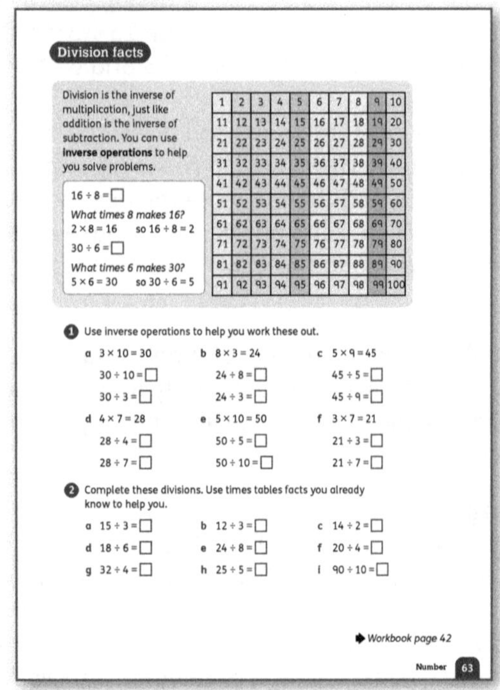

Materials

Egg boxes and fruit trays

Warm-up

Demonstrate how to use the 100-chart on **Pupil Book 3 page 63** to work out which number times 6 makes 24. Count up in sixes: 6, 12, 18, 24. Write $4 \times 6 = 24$ and the related division facts: $24 \div 6 = 4$ and $24 \div 4 = 6$. Repeat for other multiplications.

Focus

- **Pupil Book 3 page 63** gives the children plenty of extra practice in exploring multiplication and division fact families.
- Revise the term *inverse operation* and elicit that division is the inverse of multiplication. Then work through the examples with the class.
- Then let the children complete questions 1 and 2 independently.

Follow-up

Use **Workbook 3 page 42** for additional practice and consolidation of using arrays to find division facts.

Challenge

Some children may find the repetitive practice of tables boring if they have already grasped them. Ask children who would like more challenge to choose a particular fact family and illustrate it with a real-life example.

Support

Bring items into class that are arranged in the groupings that the children find difficult to recall by heart. Egg boxes are a good example of groups of 6, and it is useful for the children to learn about dozens and half-dozens. Fruit trays can be cut into fours or eights and used to practise tables and related division facts.

> ### Answers for Pupil Book 3 page 63
>
> **1** a 3, 10   b 3, 8   c 9, 5
>   d 7, 4   e 10, 5   f 7, 3
> **2** a 5   b 4   c 7
>   d 3   e 3   f 5
>   g 8   h 5   i 9

> ### Answers for Workbook 3 page 42
>
> **1** 2 rows of 6, $12 \div 2 = 6$
>   3 rows of 4, $12 \div 3 = 4$
>   4 rows of 3, $12 \div 4 = 3$
>   6 rows of 2, $12 \div 6 = 2$
> **2** 8 ways:
>   1 row of 24, $24 \div 1 = 24$
>   2 rows of 12, $24 \div 2 = 12$
>   3 rows of 8, $24 \div 3 = 8$
>   4 rows of 6, $24 \div 4 = 6$
>   6 rows of 4, $24 \div 6 = 4$
>   8 rows of 3, $24 \div 8 = 3$
>   12 rows of 2, $24 \div 12 = 2$
>   24 rows of 1, $24 \div 1 = 24$

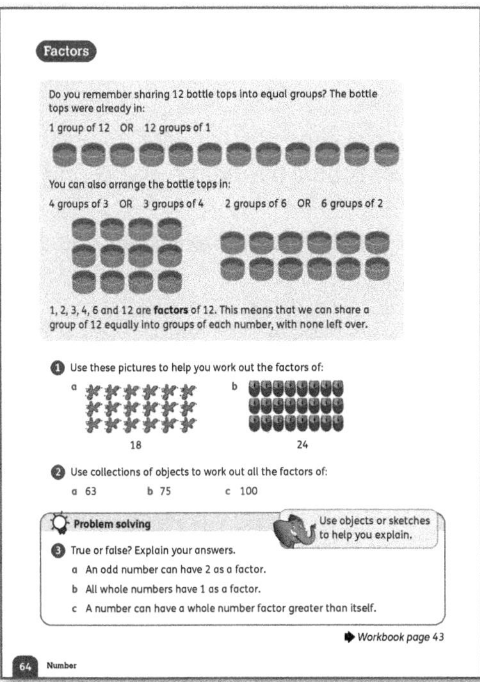

Materials
Small objects such as bottle tops, beans or counters

Warm-up
- Write $2 \times 3 = 6$ on the board. Draw arrows to the 2 and the 3 and write the word *factor*. Draw an arrow to the 6 and write the word *product*.
- Explain that when we multiply one number by another, the two numbers are called factors. The total that results when we multiply one factor by another is called the product.
- You can write several more multiplication facts on the board and have the children identify factors and products. Emphasise that 'factor × factor = product'.
- Let the children write multiplication sentences on the board and identify factors and products. Then let them draw and label some arrays in the same way.

Focus
- The only new concept in this lesson is the term factor. The children will be working with familiar multiplication facts and learning to associate the format of a multiplication sentence with the words factor and product.
- Work through the example on **Pupil Book 3 page 64** as revision, before having the children complete questions 1–3 independently.
- <u>Problem solving:</u> For question 3, ensure that the children understand that a *whole number* simply means a number that has no fractional parts.

Follow-up
Workbook 3 page 43 uses bar models to help the children visualise the division of a quantity into equal parts. Discuss the questions before the children complete them independently.

Challenge
Let the children draw their own bar models of given division facts.

Support
Provide more examples of arrays using concrete objects, and let the children identify the factors and product. Emphasise that the product is the total of the whole array. The factors are the numbers of rows and columns.

Interesting mistakes
- The terms factor and product are specific to multiplication and division.
- A factor is not the same as a fact. A factor is a number that goes into another number without a remainder.
- Similarly, although the word product has a different everyday meaning (for example, we buy and use a wide variety of products), in mathematics a product refers specifically to the result of a multiplication operation. We do not talk about the product of addition or any other operation.

Answers for Pupil Book 3 page 64
1. a 1, 2, 3, 6, 9, 18   b 1, 2, 3, 4, 6, 8, 12, 24
2. a 1, 3, 7, 9, 21, 63   b 1, 3, 5, 15, 25, 75
   c 1, 2, 4, 5, 10, 20, 25, 50, 100
3. a False. Only even numbers are divisible by 2.
   b True. Every number is divisible by 1 and itself.
   c False. The largest factor of a number is the number itself.

Answers for Workbook 3 page 43
1. a $50 + 50 = 100$ b $2 \times 50 = 100$
   c $100 \div 2 = 50$ d Half of $100 = 50$
2. a $25 + 25 = 50$ b $2 \times 25 = 50$
   c $4 \times 25 = 100$ d $25 = \frac{1}{2}$ of 50
   e $25 = \frac{1}{4}$ of 100
3. Numbers with possible explanations:

fourth row: 5	because 5 of these sections fit 1 section of 25. $25 \div 5 = 5$
fifth row: 10	because these sections are double the sections of 5. $2 \times 5 = 10$
sixth row: 20	because these sections are double the sections of 10. $2 \times 10 = 20$

Multiples

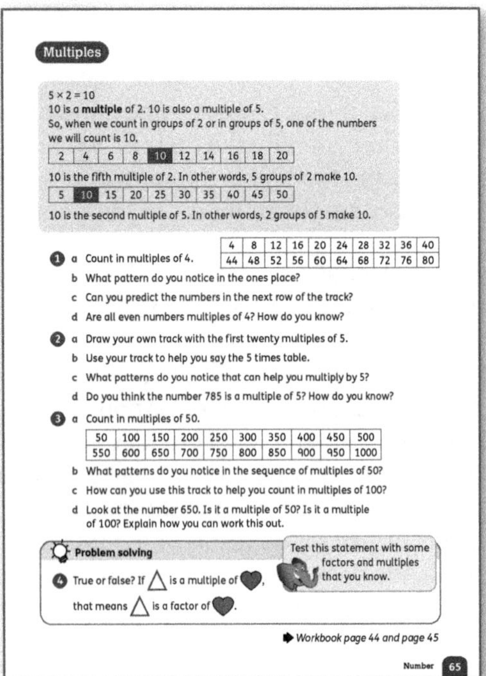

Materials
Number tracks; target diagrams (page 30)

Warm-up
- The children have used skip counting and have worked with times tables. Remind them of the term *multiple*.
- Write it on the board and ask the class: *What word does it remind you of?* (multiply)
- Explain that a multiple is the result of multiplying. *A multiple of 5 is any product of 5 and another factor.* (If necessary, revise the terms *product* and *factor*.)

Focus
- Turn to **Pupil Book 3 page 65** and read through the multiples of 2 and 5 in the examples. Ask questions such as:
  - *What is the first multiple of 2?* (2)
  - *What do we multiply 2 by to get this number?* (1)
  - *What is the fifth multiple of 2?* (10)
  - *What is the third multiple of 5?* (15)
  - *What is the next/previous multiple?* (20, 10)
- Work through questions 1 and 2 on this page before having the children complete questions 3 and 4 independently if possible.
- <u>Problem solving</u>: In question 4, the children work out whether a statement about multiples and factors is true or false.

Follow-up
- Use **Workbook 3 page 44** and **page 45** to consolidate or assess work on multiples.
- Help the children to see that the shape around each number on **Workbook 3 page 44** will help them to count the multiples. For example, the multiples of 3 have 3 sides, the multiples of 5 have 5 sides, the multiples of 8 have 8 sides, and so on.

- To complete **Workbook 3 page 45**, the children need to apply what they know about multiples of 2, 5 and 10 to be able to spot them in much greater numbers.
- Use 'Target games' (page 30) to practise multiples.

Challenge
Set the children challenges such as:
- *Find a number that is a multiple of 2, 3 and 4.* (12, 24, 36, …)
- *Find a number that is a multiple of 10 and 11.* (110, 220, 330, …)

Support
Keep relating multiples to the times tables. Explain that multiples are simply the sequence that result from a times table. If a child is finding the terminology confusing, focus on a simple times table, such as the 2 or 3 times table. The children can make their own number tracks for different times tables, by counting on.

Answers for Pupil Book 3 page 65

1 a 4, 8, 12, 16, 20, 24, 28, 32, 36, 40, 44, 48, 52, 56, 60, 64, 68, 72, 76, 80
   b 4, 8, 2, 6, 0
   c 84, 88, 92, 96, 100, 104, 108, 112, 116, 120
   d No. For example, 2 is even but not a multiple of 4.

2 a

5	10	15	20	25	30	35	40	45	50
55	60	65	70	75	80	85	90	95	100

   b
   $1 \times 5 = 5$ $\quad$ $7 \times 5 = 35$
   $2 \times 5 = 10$ $\quad$ $8 \times 5 = 40$
   $3 \times 5 = 15$ $\quad$ $9 \times 5 = 45$
   $4 \times 5 = 20$ $\quad$ $10 \times 5 = 50$
   $5 \times 5 = 25$ $\quad$ $11 \times 5 = 55$
   $6 \times 5 = 30$ $\quad$ $12 \times 5 = 60$

   c Possible answer: All the numbers end alternately in 0 and 5.
   d Yes. It ends with 5.

3 a 50, 100, 150, 200, 250, 300, 350, 400, 450, 500, …
   b Possible answer: All the numbers end alternately in 00 and 50.
   c Possible answers: Use every other number. Use the even multiples of 50.
   d Possible answer: It is a multiple of 50. It is not a multiple of 100. Multiples of 50 end in 00 or 50, and multiples of 100 end in 00.

4 True

Divisibility

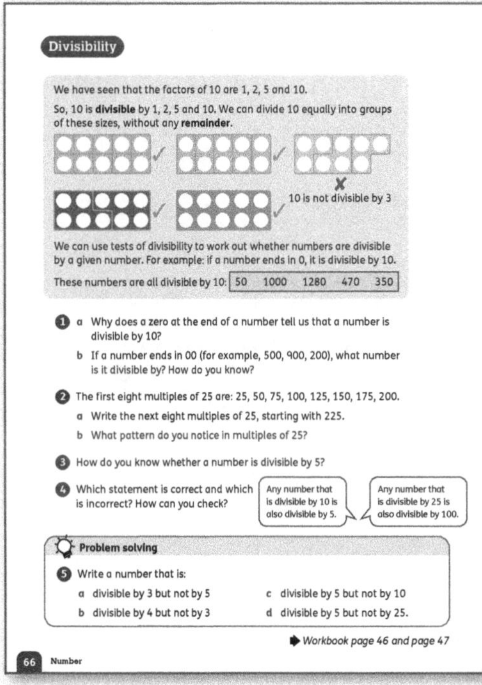

Materials

Charts of multiples or number tracks of multiples, as developed in the preceding lessons

Warm-up

- In this lesson the children continue to build on the work they have done on times tables to develop and extend their concepts and vocabulary.

- Draw a 3 × 4 array on the board. Ask: *What can we divide this number by? What groups can you see that we can divide it into without any left over?*
- Even without counting the total array, the children should be able to see that we can divide it into groups of 3 or 4. They may notice that it can also be divided into 2 groups or groups of 2.
- The children will need to apply their understanding of the ways we divide – sharing into a number of groups or sharing into groups of a given size.

Focus

- Write the number 72 on the board. Ask: *How do we know what factors 72 has? How can we work it out?* The children may notice that 72 is an even number, so 2 is a factor. Ask:
  - *What about 5? Do you think we can divide 72 by 5 without any remainder? Why or why not?* (No because 72 does not end in 0 or 5.)
  - *What about 10? Can we divide 72 into 10 equal parts? Why or why not?* (No because 72 does not end in 0.)
- Explain that when we talk about the factors of a number, we are talking about *divisibility*. If a number has 2 as a factor, we say it is *divisible* by 2. If it has 5 as a factor, we say it is divisible by 5. Write the words *divisible* and *divisibility* on the board. Ask:
  - *Who can tell me a number that is divisible by 3?*
  - *Who can tell me a number that is divisible by 4?*
  Continue with similar questions.
- Work through the examples at the top of **Pupil Book 3 page 66**. You may also wish to work through questions 1–4 together with the class, as they may generate some debate and discussion.
- <u>Problem solving:</u> For question 5, the children find numbers that satisfy more than one divisibility rule.

Follow-up

Workbook 3 page 46 is on division with 'some left over'. This activity is useful for developing the idea that if a number is not a factor, we can still divide by that number but it will leave a remainder. Let the children complete the activities on this page and discuss the results. Give the children a chance to explain how they worked out what was needed and how they worked out the answers.

Workbook 3 page 47 gives the children an opportunity to build up their own resource showing all the multiplication tables up to 10.

- You may want to let them work on this gradually over a few days, or let them work in groups on it, with each child taking responsibility for a portion of the grid.
- Give them time to check each other's work to ensure that the tables are completed.
- Go through the example of how to use counters to check the answers to make sure the children understand how to do this.
- The children should also spend some time noticing the patterns created by the multiplication sequences.

Challenge

You can adapt question 5 on **Pupil Book 3 page 66** to make it more challenging. Invite the children to write down the first 20 multiples of the first number, and then cross out the multiples that are divisible by the second number. Ask if they can find a pattern. For example, for question 5a (a number that is divisible by 3 but not by 5), they would write:

3   6   9   12   ~~15~~   18   21   24   27   ~~30~~
33   36   39   42   ~~45~~   48   51   54   57   ~~60~~

The children may notice first that they have crossed out every fifth multiple. Ask what pattern they notice if they write the sequence of crossed-out numbers. They may notice that the crossed-out numbers form the pattern of multiples of 15, which is the product of 3 and 5.

Support

The children may need support to understand the concept of divisibility. The word *divisibility* comes from the words *divide* and *ability*. It tells us if we are able to divide a number by a given factor without any left over. Set out arrays of objects and ask, for example:

- *Do you think we can divide this number into twos without any left over?*
- *How can you check?*
- *Let's try it. Are there any left over?*
- *What makes it divisible by 2?*

Interesting mistakes

The children may sometimes identify patterns that are incorrect. For example, all multiples of 100 end in a 0, but not all numbers ending in 0 are multiples of 100. Similarly, all multiples of 25 end in 5 or 0, but not all numbers ending in these digits are multiples of 25. Encourage the children to examine what they need to show to disprove a rule.

Answers for Pupil Book 3 page 66

1 **a** All multiples of 10 end in 0.
   b 100. All multiples of 100 end in 00.

2 **a** 225, 250, 275, 300, 325, 350, 375, 400
   b The last two digits follow the pattern 25, 50, 75, 00.

3 It ends in 0 or 5.

4 Correct statement: Any number that is divisible by 10 is also divisible by 5.
   Incorrect statement: Any number that is divisible by 25 is also divisible by 100.
   Possible answers: Any number that is divisible by 10 ends in 0, so that number is also divisible by 5. 50 is divisible by 25, but is not divisible by 100.

5 Possible answers:
   a 9   **b** 16   **c** 25   **d** 20

Answers for Workbook 3 page 46

1 **a** Yes, 1 left over   **b** Yes, 3 left over
   c Yes, 3 left over   **d** Yes, 1 left over
   e Yes, 3 left over   **f** Yes, 1 left over

Answers for Workbook 3 page 47

1 

×	1	2	3	4	5	6	7	8	9	10
1	1	2	3	4	5	6	7	8	9	10
2	2	4	6	8	10	12	14	16	18	20
3	3	6	9	12	15	18	21	24	27	30
4	4	8	12	16	20	24	28	32	36	40
5	5	10	15	20	25	30	35	40	45	50
6	6	12	18	24	30	36	42	48	54	60
7	7	14	21	28	35	42	49	56	63	70
8	8	16	24	32	40	48	56	64	72	80
9	9	18	27	36	45	54	63	72	81	90
10	10	20	30	40	50	60	70	80	90	100

Multiples of 10

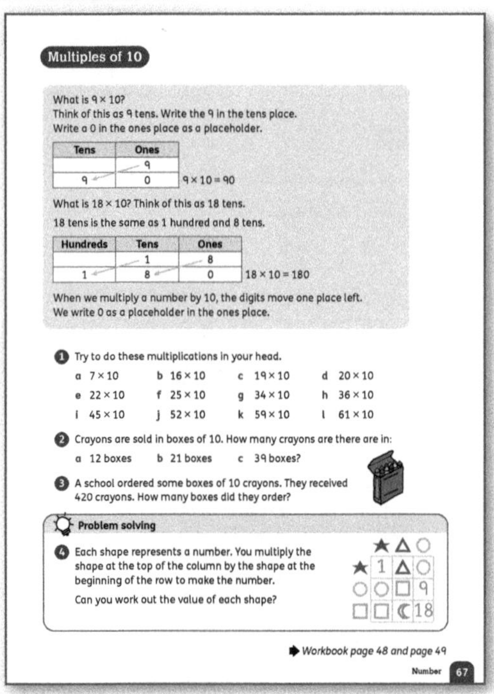

Materials

Place-value tables (page 23): interlocking cubes (page 21)

Warm-up

- Show the children three towers of 10 interlocking cubes. Count up in tens to show that 3 tens = 30. Repeat for other multiples of 10.
- Put three counters in the tens column of a place-value table and point out again that 3 tens = 30. Repeat for other multiples of 10.

Focus

- Use **Pupil Book 3 page 67** to revise multiplication by 10.
- Discuss the effect that multiplying by 10 has on a number, using the place-value tables and examples at the top of the page.
- Let the children work on their own to solve the problems in questions 1–3.
- Observe them as they work to see who can multiply by 10 mentally.
- Problem solving: Question 4 presents a multiplication grid with missing numbers represented by shapes. The children can work on this in pairs.

Follow-up

Use **Workbook 3 page 48** to consolidate multiplication patterns covered in this unit and **Workbook 3 page 49** to consolidate multiplication and division by 10.

Challenge

The children can use question 4 in the **Pupil Book** as a model and make up similar shape multiplication problems for a partner to solve.

Support

The children may need to practise counting in tens and representing numbers with an emphasis on place value. Use 'Show it four ways' (pages 27–28) to help reinforce place value.

Interesting mistakes

The children may talk about 'adding a 0' when they multiply by 10 (and parents may inadvertently reinforce this). Remind them that adding 0 doesn't change a number, whereas multiplying by 10 makes a number 10 times greater. So we don't say that we 'add 0', we say that the digits move one place to the left and we write a 0 as a placeholder.

Answers for Pupil Book 3 page 67

1 a 70 b 160 c 190 d 200
  e 220 f 250 g 340 h 360
  i 450 j 520 k 590 l 610

2 a 120 b 210 c 390

3 42

4 star = 1, triangle = 2, circle = 3, square = 6, moon = 12

Answers for Workbook 3 page 48

1 a (20, 18, 16,) 14, 12, 10, 8
  b (5, 10,) 15, 20, (25,) 30, 35
  c (10, 20, 30,) 40, 50, 60, 70
  d (3, 6, 9,) 12, 15, 18, 21
  e (12,) 14, (16, 18,) 20, 22, 24
  f (4, 8, 12,) 16, 20, 24, 28
  g (36, 32,) 28, 24, 20, 16, 12

2

×	1	2	3	4	5	6	7	8	9	10	11	12
2	2	4	6	8	10	12	14	16	18	20	22	24
3	3	6	9	12	15	18	21	24	27	30	33	36
4	4	8	12	16	20	24	28	32	36	40	44	48
5	5	10	15	20	25	30	35	40	45	50	55	60
10	10	20	30	40	50	60	70	80	90	100	110	120

3 1st column: 2, 4, 6, 8, 10
  2nd column: 20, 40, 60, 80, 100
  3rd column: 30, 50, 70, 90

Answers for Workbook 3 page 49

1 a

Number	7	4	9	2	6	8	3
× 10	70	40	90	20	60	80	30

b

Number	12	15	16	18	21	23	25
× 10	129	150	160	180	210	230	250

c

Number	20	22	26	30	31	36	39
× 10	200	220	260	300	310	360	390

d

Number	40	43	46	47	48	50	54
× 10	400	430	460	470	480	500	540

e

Number	68	74	81	88	90	95	99
× 10	680	740	810	880	900	950	990

2 a

Number	100	120	150	190	250	390	350	380	400
÷ 10	10	12	15	19	25	39	35	38	40

b

Number	1000	970	940	870	600	340	800	810	700
÷ 10	100	97	94	87	60	34	80	81	70

c

Number	200	820	50	110	140	270	760	40	150
÷ 10	20	82	5	11	14	27	76	4	15

d

Number	100	1000	200	500	900	700	600	400	300
÷ 10	1	10	2	5	9	7	6	4	3

Doubling and halving

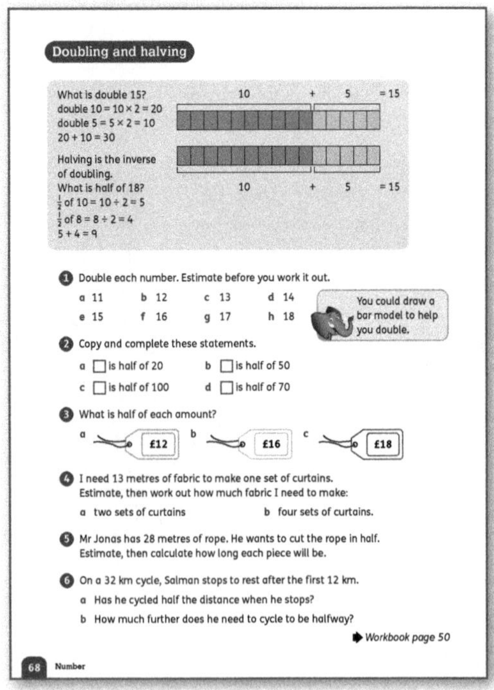

Materials
Place-value cards (pages 22–23)

Warm-up
- Remind the children how to partition numbers into tens and ones using place-value cards as necessary.
- Show them that they can use this to multiply '-teen numbers'.
- Do some examples as a class. For example:
  $3 \times 14 = 3 \times 10 + 3 \times 4 = 30 + 12 = 42$

Focus
- Revise doubling and halving using the examples at the top of **Pupil Book 3 page 68**.
- Then let the children complete questions 1–6 on their own.
- Check that they can double and halve whole numbers before moving on.

Follow-up
Workbook 3 page 50 provides further practice in doubling and halving.

Answers for Pupil Book 3 page 68
1   a 22   b 24   c 26   d 28
    e 30   f 32   g 34   h 36
2   a 10   b 25   c 50   d 35
3   a £6   b £8   c £9
4   a 26 m   b 52 m
5   estimate: 15 m   calculation: 14 m
6   a No   b 4 m

Answers for Workbook 3 page 50
1   a 24, 34, 42, 50   b 13, 15, 20, 50
    c 4, 7, 9, 25   d 6, 10, 20, 12
2   a 30, 60, 120   b 18, 9   c 12, 6, 3
3   a 50   b 100   c 300   d 400

End-of-unit check

Assess the work done in this unit by asking questions such as:
- *What is 2 times/3 times/4 times/5 times/10 times (give a number)?*
- *How many are there in 4 groups of 2? (8)*
- *What are 3 sevens? (21)*
- *Start on 16 and make 3 jumps of 4 along a number line. What numbers do you land on? (28)*
- *How many groups of 5 do you need to make 30? (6)*
- *How else can we write 4 sixes? (24. Alternative correct answers: 6 fours; 2 twelves; 12 twos)*
- *If one mango costs $6, how much do 8 mangoes cost? ($48)*
- *Six mangoes are shared equally between 3 people. How many mangoes does each person get? (2)*
- *How many sets of 5 can we make from 10? (2)*
- *There are 12 sweets in a bag. How many children will get 2 sweets each if they share the sweets equally? (6)*
- *What is the missing number? $8 \div 2 =$ ____ (4)*
- *Which of these numbers are even: 2, 5, 9, 11, 12? (2, 12)*
- *I share 6 buttons equally into 2 groups. How many buttons are there in each group? (3)*
- *I share 8 apples equally between 4 people. How many apples does each person get? (2)*
- *I want to give 5 people 2 sweets each. How many sweets do I need? (10)*
- *If I share 14 cakes equally between 4 children, how many cakes are left? (2)*
- *Can you tell me what happens when you multiply a number by 10? (It makes a number 10 times greater. The digits move one place to the left and we write a 0 as a placeholder.)*
- *What is 3 times/5 times (give any -teen number)?*
- *What is double/half of (give a number)?*

Perimeter and area

Learning objectives
- Distinguish between perimeter and area.
- Understand that perimeter is the total distance around a 2D shape and can be calculated by adding lengths.
- Understand that area is how much space a 2D shape occupies within its boundary.
- Estimate, measure and calculate the perimeter of a shape using appropriate metric units and area on a square grid.

Key words
perimeter   length   distance   sides   add
centimetre   area   square units
square centimetre   square metre

Unit introduction

Materials
Chalk, masking tape or string; metre rule or tape measure

Teaching guidance
Use any area with plenty of open space, such as the playground or the school hall. Use chalk, masking tape or string to mark some large squares of different sizes on the ground. For example, make squares measuring 2 m × 2 m, 3 m × 3 m and 5 m × 5 m. Ask:
- *What shapes do you see?* (squares)
- *Which is the largest? Which is the smallest?*

Explain to the class that they are going to measure how many steps it takes to go all the way around each square. Invite a child to start at one corner of the largest square. Let the class count the child's steps as they walk along one side. Ask:
- *Who can guess how many steps it will take for* [child's name] *to walk the whole way around the square?*
- *How did you work it out?*
- *Who agrees? Who has a different idea?*

Let the child continue walking along the next side, then the next, then the next.

Point at one of the other squares and say: *Will we need to take more steps or fewer steps to walk around the outside of this square? Why?*

Let the children measure how many steps it takes to walk around each square. The answers may vary as the children are using paces, which are non-standard measures. If you wish, you can raise this by asking: *Why do you think a different child gets a different number? What unit can we use to measure it?*

Demonstrate measuring the *distance* around each square with a metre rule or tape measure. Then let the children try this.

Back in the classroom, explain that the distance around a shape is called the *perimeter*. Write the word perimeter on the board. Explain that *peri-* means around, and *-meter* is the unit we use to measure *length*.

Perimeter

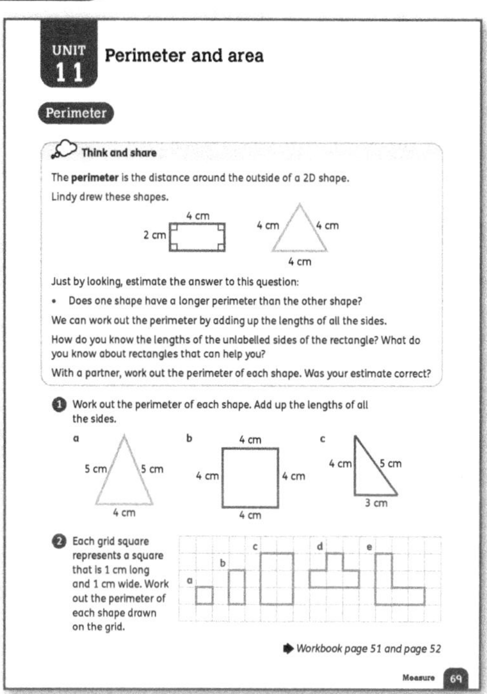

Materials
Rulers; centimetre-squared paper

Warm-up
Draw various polygons (rectangles, squares, triangles) on the board, and let the children guess and then measure the perimeter of each.

Focus
- <u>Think and share:</u> Look at the shapes at the top of **Pupil Book 3 page 69**. Ask the children to say, just by looking, whether they think one of the shapes has a longer perimeter than the other.
- Explain that sometimes we draw a sketch of a shape. A sketch does not have exactly the right measurements, but we label the *sides* with the correct measurements. We *add* the measurements to find the perimeter.
- Ask the children to think about how to check the perimeter of the shapes and then let them do it. When they have finished, listen to the various methods that the children used. For example, they may have added up all the sides. For the rectangle, they may have multiplied 2 cm by 2 and 4 cm by 2, then added the products. For the triangle, they may have multiplied 4 cm by 3.
- The children can work independently through questions 1 and 2.

Follow-up

Use **Workbook 3 page 51** and **page 52** to consolidate the work on perimeter. Note that on **Workbook 3 page 51** the children measure the perimeter to the nearest *centimetre* using their rulers. On **Workbook 3 page 52**, they calculate perimeter using the given lengths.

Challenge

Give the children centimetre-squared paper and set this challenge: *Draw an irregular shape with as many corners as you can, with a perimeter of 30 cm.* Then let the children compare their shapes and see whose shape has the most corners.

Support

Demonstrate again, if necessary, how to use a ruler to measure lengths in centimetres. Have the children draw a line of a given length. Then help them to construct more sides to sketch a square or rectangle. Explain that perimeter is the lengths of all the sides added up.

Interesting mistakes

The children may confuse height and width with perimeter. Where these mistakes occur, let them compare answers with a partner and compare how they reached their answer. Ask: *What were you trying to work out? How do we work out perimeter?*

Some children may also be confused by sketches that are not drawn to scale. Emphasise that when measurements are labelled, the children should use the given measurements, not measure the sides.

Answers for Pupil book 3 page 69

<u>Think and share:</u> No – they both have the same perimeter.
Opposite sides of a rectangles are the same length. The perimeter of both shapes is 12 cm.

1 a 14 cm   b 16 cm   c 12 cm
2 a 4 cm   b 6 cm   c 10 cm
   d 10 cm   e 12 cm

Answers for Workbook 3 page 51

1 a 3 + 3 + 3 + 3 = 12 Perimeter: 12 cm
   b 3 + 3 + 4 = 10 Perimeter: 10 cm
   c 2 + 2 + 2 + 2 + 2 = 10 Perimeter: 10 cm
   d 3 + 5 + 3 + 5 = 16 Perimeter: 16 cm
   e 2 + 4 + 5 = 11 Perimeter: 11 cm
   f 2 + 2 + 2 + 2 + 2 = 10 Perimeter: 10 cm

Answers for Workbook 3 page 52

1 a 4 + 32 + 4 + 32 = 72 Perimeter = 72 cm
   b $\frac{1}{2} + 1 + \frac{1}{2} + 1 = 3$ Perimeter = 3 m
   c 3 + 2 + 3 + 2 = 10 Perimeter = 10 m
   d 20 × 3 = 60 Perimeter = 60 cm
2 30 cm

Area

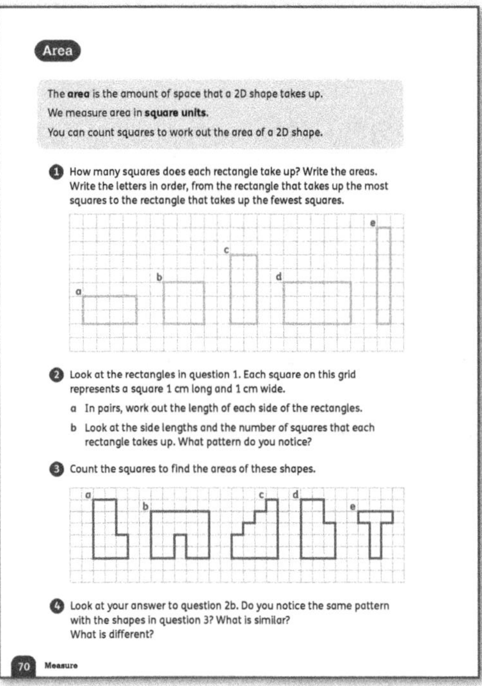

Materials

Centimetre-squared paper

Warm-up

- Draw a 3 cm × 5 cm rectangle on centimetre-squared paper. Show it to the class and ask what they see.
- Explain that each side of a square is 1 centimetre long. *We call these* square centimetres *because each side is equal to 1 centimetre.*
- Now explain that we can use the squares to tell us how much space the shape takes up. Ask: *How many squares does this shape take up on the grid?* Let them count the squares to work it out.
- Explain that the amount of flat space that a shape takes up is called its *area*.

Focus

- Tell the children that we use special units called *square units* to measure area.
- Demonstrate how we write the notation for square centimetres and how to write the superscript ^{2} after the unit of length: 1 square centimetre = 1 cm^2.
- Turn to **Pupil Book 3 page 70**, and work through the information at the top of the page with the class.
- You can work through question 1 together with the class and let the children complete questions 2–4 independently.

Challenge

Ask the children to try to find a way to work out the area of a triangle.

Support

If necessary, revise the units of measure and show the children centimetres on a ruler. Let them identify things that are about 1 cm (the width of a finger, thickness of a marker pen, etc.). Make sure they understand what a centimetre is and how to use a ruler.

1   **a** 8   **b** 9   **c** 10   **d** 15   **e** 7
    In order: d, c, b, a, e

2   **a** a: 2, 4; b: 3, 3; c: 5, 2; d: 3, 5; e: 7, 1
   b The number of squares is the two lengths multiplied together.

3   **a** 12   **b** 18   **c** 14   **d** 13   **e** 6

4   No, the same pattern is not in the shapes in question 3.
    Possible answer: All the shapes have straight edges and right angles, but the shapes in question 1 are rectangles and the shapes in question 3 are not.

Work with perimeter and area

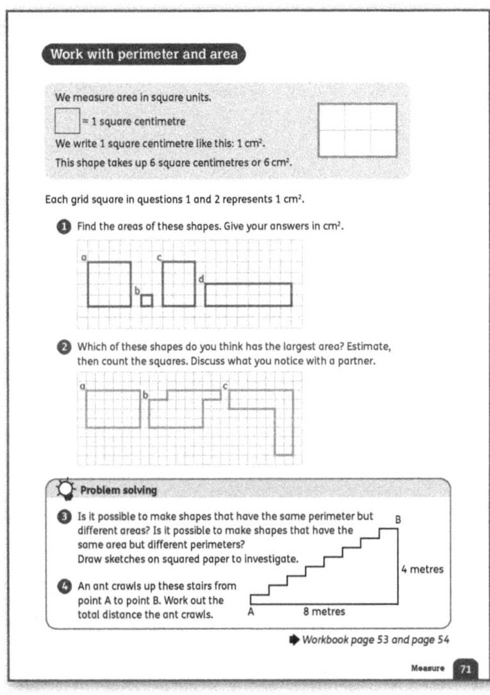

Materials

Centimetre-squared paper, adhesive paper or contact paper in at least two different colours, scissors; rulers

Warm-up

- As a practical activity, give the children centimetre-squared paper and contact paper (sticky paper with a removeable backing) in a variety of colours. Let them use the contact paper to create designs. They then work out what area they covered in each colour.
- If you wish, pre-cut the contact paper into centimetre squares so that this is also a tiling exercise.

Focus

- Work through the information at the top of **Pupil Book 3 page 71** and remind the children about the standard units for measuring area (square centimetre and *square metre*) and how to write the units correctly with the superscript ^{2} after the unit of length.

- Ask the children if they can think of places we might measure using square metres rather than square centimetres. Examples include:
  - a floor (to calculate how much carpet is needed or how many tiles are needed)
  - a patio (how many tiles are needed)
  - a wall (how many pots of paint are needed).
- Begin working through questions 1 and 2 on **Pupil Book 3 page 71** with the children, then let them complete the work independently if possible.
- <u>Problem solving:</u> The children can discuss question 3 in pairs and draw sketches as they investigate.
- Let the children discuss question 4 in pairs or groups to try and work it out. If necessary, give them hints:
  - *The stairs are at a right angle.*
  - *The ant is going straight up a bit, then straight across a bit, then straight up again …*
  - *How many will all the 'up' or vertical bits add up to? (4 m) How many will all the 'across' or horizontal bits add up to? (8 m)*

Follow-up

Workbook 3 page 53 and **page 54** provide practice in measuring areas. Space is also provided for the children to draw their own shapes and work out the perimeter and area of the shapes. Some children may wish to explore a variety of irregular shapes, whereas others may prefer to explore a variety of squares and rectangles.

Challenge

The children can draw more shapes on squared paper and estimate then measure their area.

1   **a** 16 cm²   **b** 1 cm²
   c 12 cm²   **d** 16 cm²

2   Individual answers for the estimates
   a 24 cm²   **b** 24 cm²   **c** 24 cm²

3   Yes. Individual answers   **4** 12 m

1   **a** 4 cm²   **b** 3 cm²   **c** 7 cm²
   d 5 cm²   **e** 15 cm²

2   Individual answers

1   Individual answers

Assess the work done on perimeter and area by asking questions such as:

- *Which of these definitions of the perimeter of a shape are correct? The perimeter is:*
  - how long the shape is (incorrect)
  - the distance around the outside of the shape (correct)
  - the length of one of the sides (incorrect)
  - the sum of the lengths of all the sides (correct)
  - the difference between one side and another. (incorrect)

Give the children shapes with the side measurements labelled. For each shape ask: *What is the perimeter of this shape?*

Give the children shapes drawn on centimetre-squared paper and ask: *What is the perimeter? What are the units for this perimeter?* (cm) *What is the area? What are the units for this area?* (cm^2)

Give the children shapes drawn on plain paper and ask them to measure the sides and calculate the perimeters.

Give the children centimetre-squared paper and ask them to draw their own shapes with a given area. As a challenge, they can try to draw different shapes with the same area.

UNIT 12 Data

Learning objectives

- Plan and conduct an investigation.
- Ask and answer non-statistical and statistical questions.
- Record, organise and represent categorical and discrete data.
- Choose and explain which representation to use in a given situation: Venn and Carroll diagrams; tally and frequency tables; pictograms and bar charts.
- Interpret and present data using bar charts, pictograms and tables.
- Interpret data, identifying similarities and variations, within and between data sets, to answer non-statistical and statistical questions.
- Discuss predictions and conclusions.
- Solve one-step and two-step questions using information presented in scaled bar charts, pictograms and tables.

Key words

table   row   column   tally mark   tally table
survey   bar chart   total   Venn diagram   element
overlap   intersection   Carroll diagram   category

Teaching guidance

Here are two ideas for practical introductory activities for this unit:

- Prepare a large table, like the one below, either for the whole class or for groups of children. Include as many rows as there are children in the class or group. You could change some of the headings if you wish.
- Let the children complete the table – each child fills in one row of the chart with their own personal data. Spend time examining the data in the table to find similarities and differences, while teaching the words *table*, *row* and *column*.

Name	Birth month	Age in years	Eye colour	Number of siblings

- Ask the children to collect examples of tables from magazines, newspapers or the Internet. Use these to make a class display. Focus on the idea that a table presents information in rows and columns and that the heading of each column tells you what is in that column. Do not worry too much about the actual content of the tables, as much of it will be at too high a level for the children.

Unit introduction

Materials

Large table as shown in the Teaching guidance below; tables from magazines, newspapers or the Internet

Charts and tables

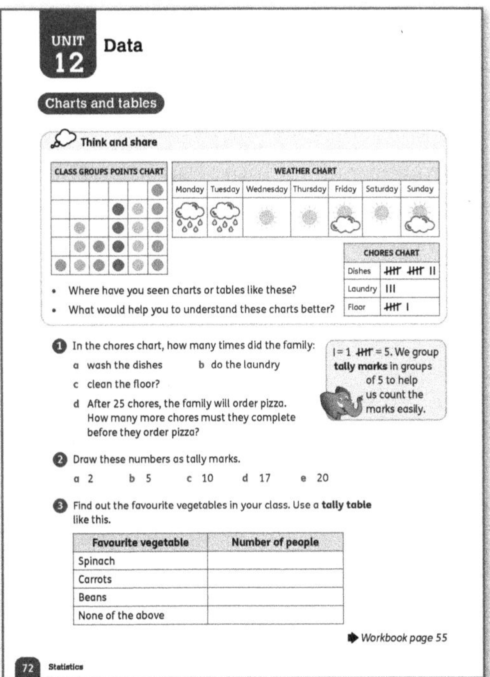

Warm-up

<u>Think and share:</u> Discuss the three charts in this section of **Pupil Book 3 page 72**. Ask:

- *What do you notice about the first chart?*
- *Which way do the lines go?*
- *How are the dots arranged?*
- *What do you think this chart shows?*
- *What do you think each dot could be telling us?*
- *What would help us to understand it better?*
- *What is a weather chart?*
- *What about the chores chart – what do you think it is showing?*
- *What chores are listed?*
- *How many times did the family do each chore?*
- *Where else do you see charts?*
- *When are charts useful? Do you ever use charts at home? What for?*

The idea here is not to teach the specific content of the charts, but to get the children thinking about the way the information is arranged, what is shown and what is left out, and why we might arrange information in this way.

Focus

- In question 1 on **Pupil Book 3 page 72**, the children discuss the *tally marks* in the chores chart. They have encountered tally marks before, but you may need to revise the way we write tally marks – one stroke for each item up to four, then a line through four strokes to make a group of five. This makes it easier to count the marks in fives.
- The children practise writing tally marks in question 2.
- The children collect information from their classmates about who prefers which vegetables in question 3.

Follow-up

Use **Workbook 3 page 55** to consolidate using tally marks and *tally tables* in the context of number bonds. This page focuses on bonds to 20 and then on how to adapt these bonds when working with numbers that are one more or one less. For example, if the known fact is 10 + 10 = 20, we can work out that 10 + 9 = 19.

Support

When working with tables, some children may have difficulty remembering which are the columns and which are the rows. When you make and display tables, you may find it useful to add large labels and arrows to indicate the columns and the rows.

Answers for Pupil Book 3 page 72

<u>Think and share:</u> Possible answers: on the weather forecast, on a website, in a maths book, in a magazine in the classroom

Individual answers

1 a 12   b 3   c 6   d 4

2 a ||
b ||||
c |||| ||||
d |||| |||| |||| ||
e |||| |||| |||| |||| ||||

3 Individual answers

Answers for Workbook 3 page 55

1 |||| |||| |||| |||| ||||
There are 24 twenties. All the numbers in the box add up to 480.

2 Individual answers. For example:
12 + 7 = 19   19 − 12 = 7   19 − 7 = 12
1 + 18 = 19   19 − 1 = 18   19 − 18 = 1

Read tables

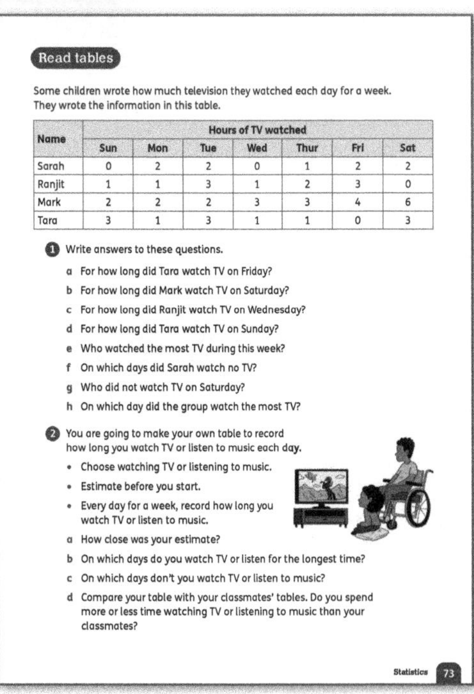

Name	Hours of TV watched						
	Sun	Mon	Tue	Wed	Thur	Fri	Sat
Sarah	0	2	2	0	1	2	2
Ranjit	1	1	3	1	2	3	0
Mark	2	2	2	3	3	4	6
Tara	3	1	3	1	1	0	3

Warm-up

- Draw a table on the board with three columns with the headings Mon, Tues, Wed, and three or four rows for children's names.
- Ask the children how much time they think they spent playing with friends on each of those days this week. Then record the names and times for three or four children in the table.

Focus

- Turn to **Pupil Book 3 page 73**.
- Work through some or all of the questions in question 1 with the class to make sure the children can read the table before they find and write the answers.
- Work through question 2 as a class orally before letting the children complete the activity on their own. Give the children some time to draw up their own table. They will need to record how long they spend watching TV or listening to music every day for the week before you return to this as a class.

Challenge

Give the children some information tables from newspapers, magazines or the Internet. Ask the children to make up questions about one of the tables, then exchange them with a partner.

Support

Show the children how to use a ruler or another straight edge to help them read along a row or down a column.

Answers for Pupil Book 3 page 73

1
  a 0 hours   b 6 hours   c 1 hour
  d 3 hours   e Mark
  f Sunday and Wednesday
  g Ranjit   h Saturday

2 Individual answers

Pictograms

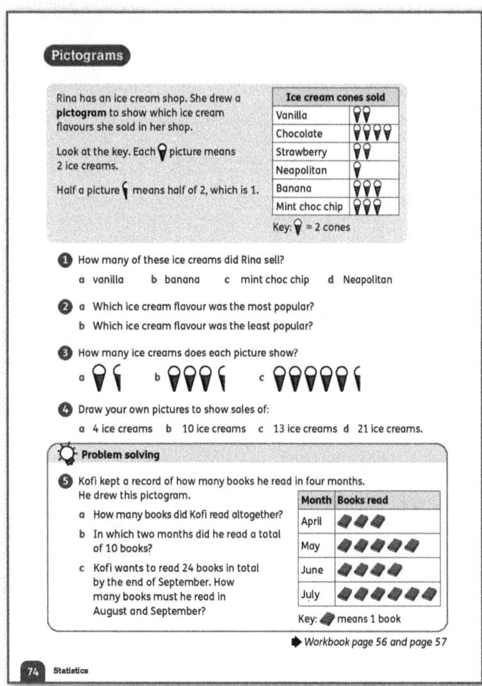

Materials

Cubes (or counters)

Warm-up

- The children could devise and carry out a *survey* of ten of their classmates. For example, they could ask each classmate to choose their favourite drink from a choice of four. The children could present their results using cubes, with one cube representing one child.
- The children could practise recording the results of a survey in a tally table.

Focus

- Work through **Pupil Book 3 page 74** to give the children practice in answering questions about a given pictogram. Once you have worked through the information in the box at the top of the page, let the children work independently to complete questions 1–4.
- Problem solving: In question 5 the children answer questions using information in a pictogram.

Follow-up

Workbook 3 page 56 provides further practice in interpreting and drawing symbols to represent different numbers. On **Workbook 3 page 57**, the children carry out their own survey, record their results in a tally chart and draw their own pictogram to show the results.

Challenge

Give pictograms with different keys, for example a circle represents 4 items. Let the children work out how to represent multiples of 4, then how to represent 1, 2 or 3 items.

Support

Previously, the children have worked with symbols that represent 1 item. When a symbol represents 2 items, half the symbol represents 1 item. You may need to create simple pictograms, similar to those in the **Pupil Book**, to provide extra practice in reading and interpreting symbols that represent 2 items. If necessary, spend some time practising doubling and halving to 20.

Answers for Pupil Book 3 page 74

1 a 4   b 6   c 6   d 2
2 a Chocolate   b Neapolitan
3 a 3   b 7   c 11
4 a 2 ice creams   b 5 ice creams
  c $6\frac{1}{2}$ ice creams   d $10\frac{1}{2}$ ice creams
5 a 18   b June and July   c 6

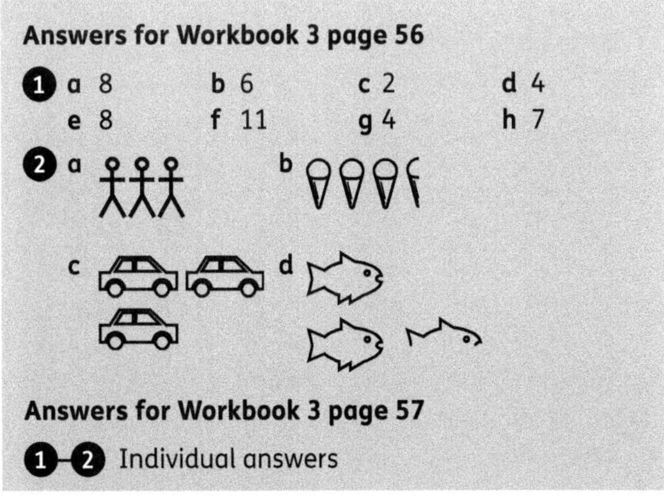

Answers for Workbook 3 page 56

1 a 8 b 6 c 2 d 4
  e 8 f 11 g 4 h 7

2 a 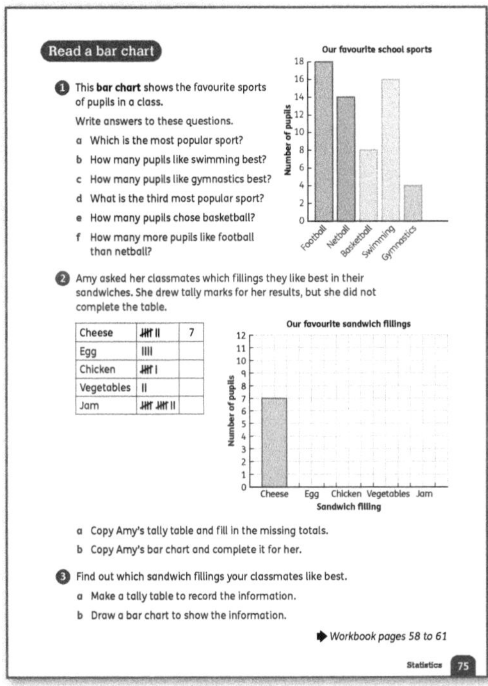 b

  c d

Answers for Workbook 3 page 57

1–**2** Individual answers

Read a bar chart

[Pupil Book 3 page 75 reproduction]

Read a bar chart

1 This **bar chart** shows the favourite sports of pupils in a class.
Write answers to these questions.
a Which is the most popular sport?
b How many pupils like swimming best?
c How many pupils like gymnastics best?
d What is the third most popular sport?
e How many pupils chose basketball?
f How many more pupils like football than netball?

2 Amy asked her classmates which fillings they like best in their sandwiches. She drew tally marks for her results, but she did not complete the table.

Cheese	𝄪 II	7
Egg	IIII	
Chicken	𝄪 I	
Vegetables	II	
Jam	𝄪 𝄪 II	

a Copy Amy's tally table and fill in the missing totals.
b Copy Amy's bar chart and complete it for her.

3 Find out which sandwich fillings your classmates like best.
a Make a tally table to record the information.
b Draw a bar chart to show the information.

➡ Workbook pages 58 to 61

Statistics 75

Materials
A variety of shapes to sort, sticky notes, squared paper, rulers

Warm-up
- Introduce the 'Our favourite school sports' *bar chart* on **Pupil Book 3 page 75**. Ask:
  ○ What kind of graph is this?
  ○ What does it show us?
  ○ What do you notice about the coloured bars?
- Explain that this is called a *bar chart*. It is similar to the block diagrams that the children have worked with previously. However, in bar charts, the bars do not touch each other and they do not show the blocks.
- Work through question 1 with the children, demonstrating how we read information from this kind of chart.

Focus
- Before moving on to question 2, demonstrate how to draw a bar chart. Demonstrate how the process is different from drawing a block diagram. The children can start by building up a block diagram, and then use this to help them construct a bar chart.
- Here are a block diagram and a bar chart that show the same data about the numbers of different shapes:

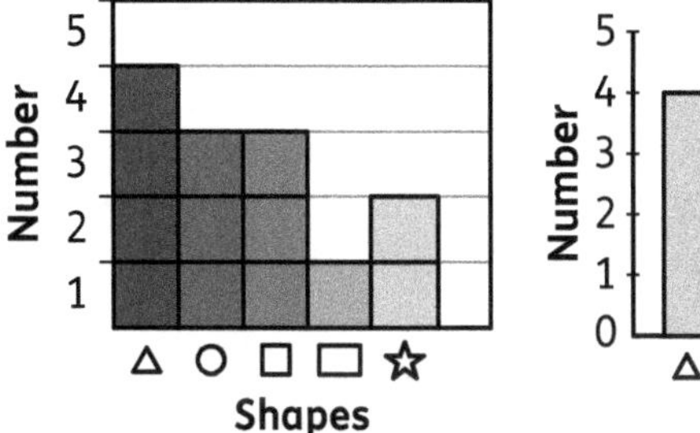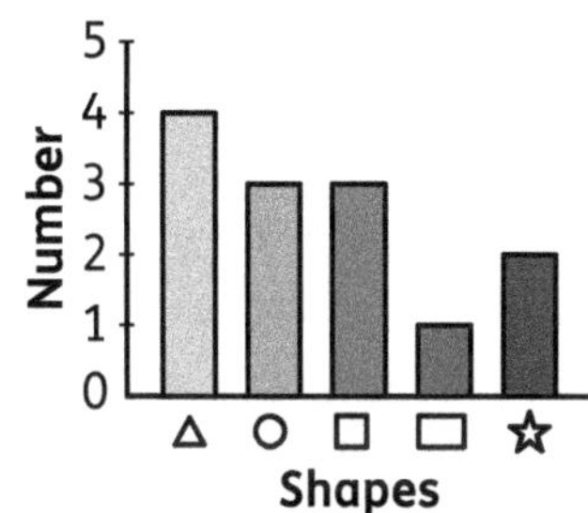

- Give each group of children a mixed set of shapes to sort. There should be uneven numbers of four or five different kinds of shapes (for example, one triangle, three circles, five rectangles and one square).
- First, the children make a list of the number of each shape or use tally marks to record them. Then they use sticky notes to build up a block diagram.
- Next, they draw a bar chart on squared paper. Demonstrate step by step how to do this:
  ○ Start by drawing a horizontal axis. This is the line that all the bars will start on.
  ○ Next, draw the vertical axis. Explain that each of the lines you have drawn is called an axis (plural axes). We can think of an axis as the line along which we track or plot our data.
  ○ Mark equal divisions on the vertical axis, showing where the numbers will be marked.
  ○ Mark equal divisions on the horizontal axis for the number of bars needed.
  ○ For each bar, lightly mark the height it will reach, reading across from the vertical axis.
  ○ Use a ruler to construct the bars.
  ○ Ensure that the axes are labelled correctly.
  ○ Finally, ensure that the graph has a title.
- In question 2, the children discuss the information in the tally table and how Amy has represented the first row as a bar in her bar chart. Let them copy the tally table and write the missing *totals*. Supervise them as they copy and complete Amy's bar chart.
- The children collect data about their classmates' favourite sandwich fillings in question 3. The children can use **Workbook 3 page 60** to record the data and draw a bar chart, or they can draw their own table and chart.

Follow-up

Workbook 3 page 58 gives the children plenty of practice in recording data using tally tables and sorting a data set using Carroll diagrams. You can use **Workbook 3 page 59** and **page 61** to provide consolidation, extra practice or assessment of drawing and interpreting bar charts.

Support

The children may need plenty of help with learning to construct their own bar charts. It is important that they understand how to draw lines with a ruler and how to mark the divisions correctly. Demonstrate how to read the number that each bar represents, by laying a ruler horizontally across the top of the bar to the vertical axis.

Answers for Pupil Book 3 page 75

1 a Football   b 16   c 4
   d Netball   e 8   f 4

2 a 4, 6, 2, 12
   b

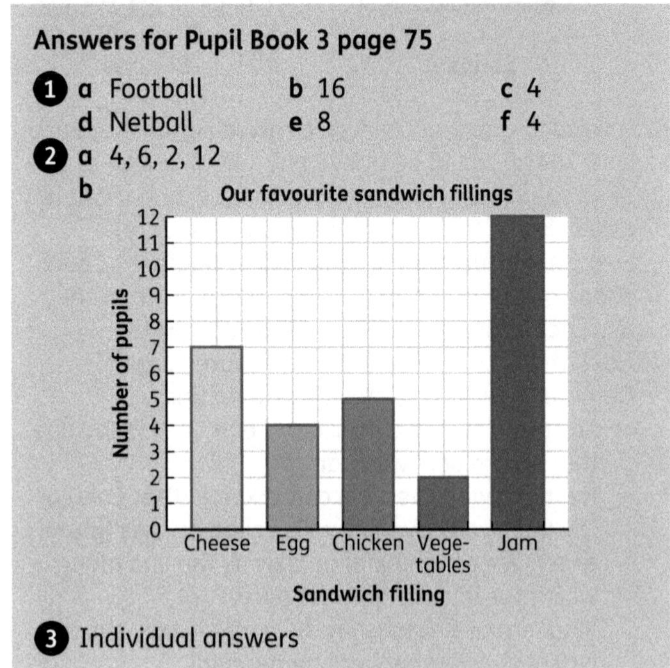

3 Individual answers

Answers for Workbook 3 page 58

1 a 12
   b Wearing shorts IIII
    Wearing a skirt IIII
    Wearing long trousers IIII
   c Wearing glasses IIII I
    Not wearing glasses IIII I
   d

	Wearing a skirt	Not wearing a skirt
Wearing glasses	1	5
Not wearing glasses	3	3

2 Individual answers

Answers for Workbook 3 page 59

1

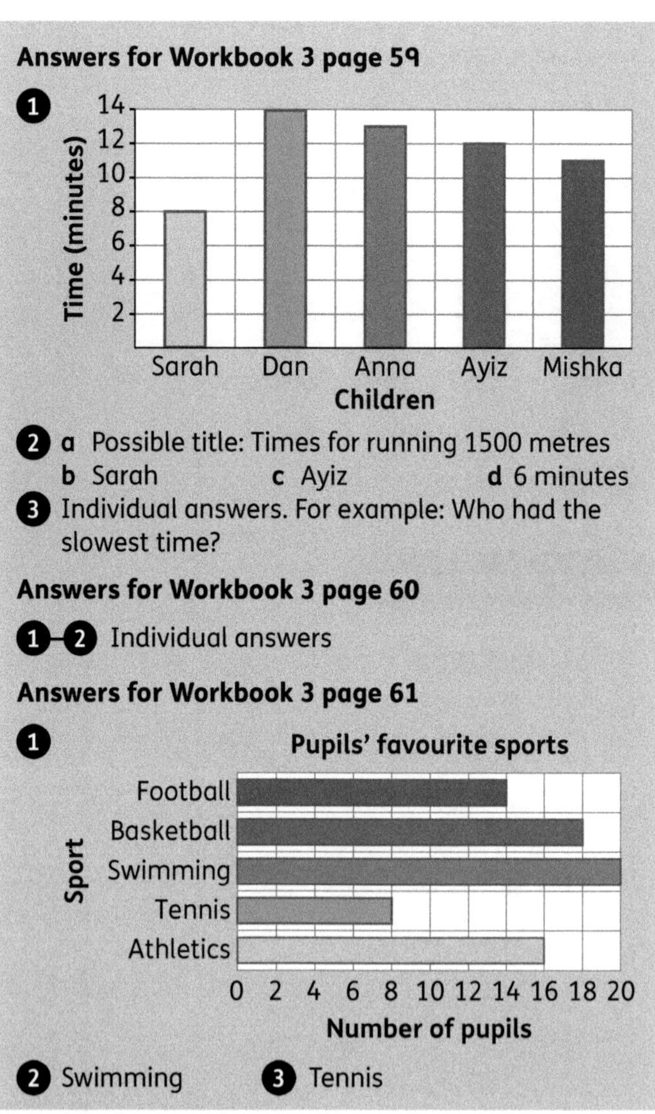

2 a Possible title: Times for running 1500 metres
   b Sarah   c Ayiz   d 6 minutes

3 Individual answers. For example: Who had the slowest time?

Answers for Workbook 3 page 60

1–**2** Individual answers

Answers for Workbook 3 page 61

1

2 Swimming   **3** Tennis

Venn diagrams

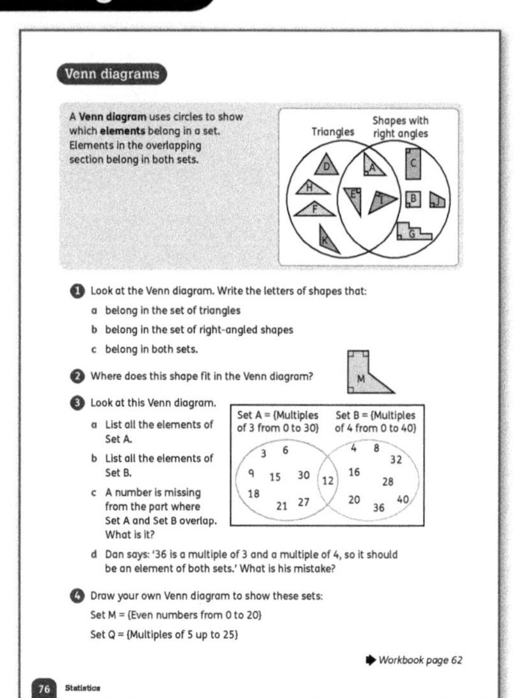

Materials
Rope or chalk to mark large circles

Warm-up
- Use rope or chalk to mark out two large intersecting circles in the school playground or hall.
- Decide on two criteria for sorting the children, such as those wearing red and those wearing blue (you will need to decide on criteria that suit your class).
- Sort the children into each circle depending on what colours they are wearing.
- Let the children work out what criteria you are using to sort them.
- Discuss which children should go into the intersection (those wearing both red and blue) and which children should stay outside the circles (those wearing neither red nor blue).
- Let the children draw their own diagrams to represent the sorting.

Here are some other ideas for sorting criteria for the intersecting circles:
- children with long hair, children with glasses
- juice bottles (or lunch boxes), things that are black or white
- things you can write or draw with, yellow things.

Focus
Work through the example at the top of **Pupil Book 3 page 76**. Ask:
- *How many shapes do you see?* (11)
- *Which two sets have been created?* (Triangles and Shapes with right angles)
- *What do you notice about all the shapes in this part (pointing to the 'Triangles' section)? What about this part (pointing to the intersection)? What about this part (pointing to the 'Shapes with right angles' section)?*
- Remind the children of the term *Venn diagram* for this method of sorting sets of items (such as objects, shapes or numbers). Explain that the items in a set are called *elements*. The correct word for the *overlap* between the sets is the *intersection*. The children do not need to know this word, but it is helpful to use the correct language from the start, so that they will recognise it when they learn it later.
- By now the children will be able to identify triangles and identify whether shapes do or do not have right angles.
- Work through question 1 with the class. For parts a and b, make sure the children understand that each set includes everything inside the circle, including the intersection. So, the set of triangles is everything inside the circle labelled 'Triangles', including the three triangles that also belong in the set of 'Shapes with right angles'.

- Continue working with the children to complete questions 2 and 3. Explain that:
  - We use curly brackets to show the elements of a set.
  - We often name sets with letters, such as Set A and Set B.
- Let the children work independently on question 4, providing support where necessary.

Follow-up
Use **Workbook 3 page 62** to consolidate or assess the children's work on Venn diagrams.

Challenge
Some children may enjoy using a Venn diagram with three circles. For example, they could draw three intersecting circles to sort:

Set A = {Animals that jump}

Set B = {Animals that swim}

Set C = {Mammals}.

Ask them to write one element in each section of the diagram.

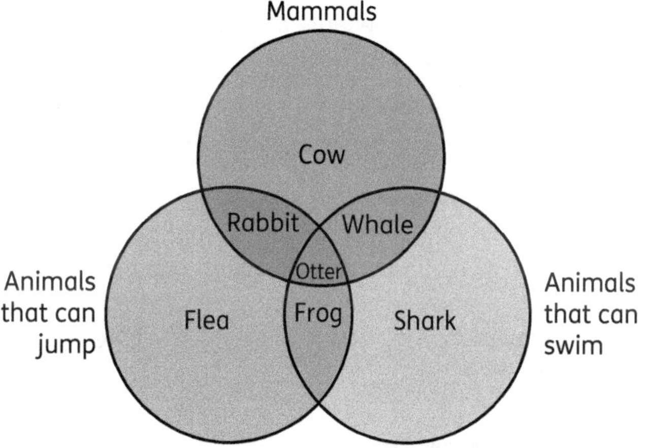

Support
Some children may need practice sorting objects into a single circle before moving on to two sets with an intersection. You can draw a circle and label it 'Green objects'. Then give the children a variety of objects in various colours. Say: *Only green objects belong in this set. Which objects should go inside the circle? Which objects do not belong in the set?*

Interesting mistakes
The children are likely to count the elements in the intersection twice, or to leave this section out altogether. Remind them to view each set as the whole circle, including the elements that are in the intersection.

Answers for Pupil Book 3 page 76

1 **a** A, D, E, F, H, I, K
  b A, B, C, E, G, I, J
  c A, E, I
2 In the circle for 'Shapes with right angles' but not in the overlapping section
3 **a** 3, 6, 9, 12, 15, 18, 21, 30
  b 4, 8, 12, 16, 20, 28, 32, 36, 40
  c 24
  d Set A only goes up to 30 and 36 is greater than 30.
4 Venn diagram drawn as follows:
  in 'Set M' circle only: 2, 4, 6, 8, 12, 14, 16, 18
  in 'Set Q' circle only: 5, 15, 25
  in the overlapping section: 10, 20

Answers for Workbook 3 page 62

1 Venn diagram completed as follows:
  in 'Underlined numbers' circle only: 29, 35, 15
  in 'Even numbers' circle only: 22, 18, 72, 20, 32
  in the overlapping section: 48, 24, 12
  outside both circles: 63, 45, 27, 51, 21

2 banana, sun, cheese, lemon, sunflower: yellow; chocolate: brown; broccoli: green; orange: orange
  Venn diagram completed as follows:
  in 'Yellow things' circle only: sun, sunflower
  in 'Things we can eat' circle only: chocolate, broccoli, orange
  in the overlapping section: banana, cheese, lemon

Carroll diagrams

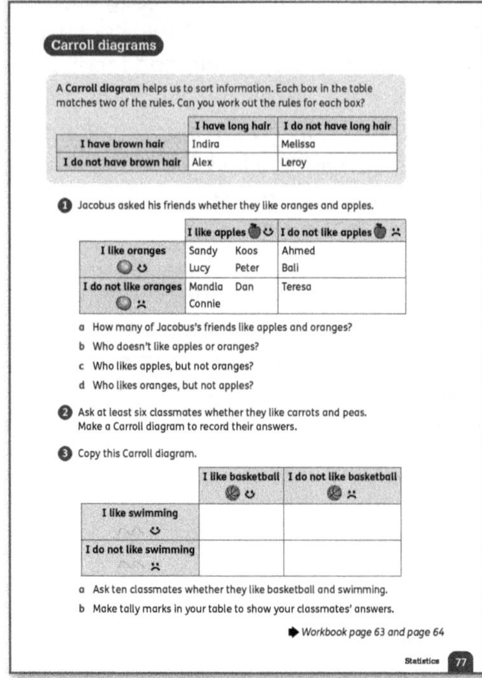

Materials

Blank Carroll diagrams; a variety of objects for the children to sort

Warm-up

- The children have used Carroll diagrams to sort data in previous years, as well as in an activity in Unit 6, but this may be the first time they have learnt the name *Carroll diagram*.
- Prepare a large blank Carroll diagram for each group. A Carroll diagram is a table with two columns and two rows, plus headings for the columns and rows. Items are sorted according to whether they do or do not have a given property. For example:

	Cube	Not a cube
Black		
Not black		

- Give the children around ten objects to sort. For example, give them ten 3D objects labelled with letters A to J, in a variety of colours and shapes. Let them sort these correctly into the blocks on their template. They can write the letter of each object in the correct block on the diagram.
- Note that you may choose other objects and *categories*. Here are some more ideas:
  - animals: flies, does not fly; has feathers, does not have feathers
  - household objects: uses electricity, does not use electricity; in the kitchen, not in the kitchen
  - clothing: blue, not blue; in pairs, not in pairs.

Focus

- Work through the explanation on **Pupil Book 3 page 77** orally as a class before the children complete questions 1–3 in pairs.
- Give them some time to carry out the survey in question 2 and to record their results.
- Have a discussion as a class about what the results of the children's different surveys show.

Follow-up

Use **Workbook 3 page 63** to provide further practice in interpreting and creating Carroll diagrams.

This year, the children need to carry out their own investigation to answer a real-life question, for example collecting data about populations (numbers) of mini-beasts. You can use **Workbook 3 page 64** to introduce the investigation and to provide a structure for the children to record and interpret their data.

However, if it is not possible or desirable for you to investigate this particular topic, you may prefer to give the children one of the following investigations:

- *Investigate which areas of the school are most popular with the children during playtime.*
- *Investigate which day of the week your classmates like best.*
- *Investigate which is the favourite colour of the children in your class.*

If you are choosing one of these investigations (or another one of your own), you will need to develop a sheet similar to the one on **Workbook 3 page 64**.

Mini-beasts are small living creatures without a backbone (invertebrates). The name *mini-beasts* is used because it allows us to talk about all these creatures without needing to use all the individual terms (insects, arachnids, worms, etc.). If the children are interested in this topic, you may like to extend it and do some classification activities.

When you do an investigation, you will need to give clear instructions and spend some time identifying suitable locations and preparing sheets for observation and recording. You will also need to make sure the children keep the data they collect (or keep it for them) so that they can use it later in the year to draw graphs.

Challenge

Explain to the children that a Carroll diagram can help us to make decisions. An action may have positive or negative consequences. We can consider these using a Carroll diagram.

Give a simple and fun example: *I have a party at home and there is a big cake. After my party, half the cake is left over. I am trying to decide whether I should bring it to school to share with everyone or keep it at home. I draw this Carroll diagram.*

	Take to school	Not take to school
Things I like about this		
Things I don't like		

Discuss with the class the possible positive and negative consequences of each decision. Here are some examples:

	Take to school	Not take to school
Things I like about this	Everyone can enjoy some cake. I enjoy sharing.	I can have more cake tomorrow.
Things I don't like	The cake will be finished.	The cake may go stale. I may eat too much and feel sick!

Ask the children which decision they would make, based on the diagram.

If the children have ideas of other decision situations, they can use Carroll diagrams to analyse the positive and negative consequences.

Support

Young children may struggle with concepts such as 'triangle' and 'not triangle' (or 'fruit' and 'not a fruit'). Make sure you give them plenty of oral practice in identifying items that are (one category) and that are not (one category). For example, you could hold up a range of writing materials and get the children to say whether each item is a pencil or not a pencil.

Interesting mistakes

In a Carroll diagram, the children may want to use four different categories instead of just two. Spend time making sure the children understand that a Carroll diagram represents the four-way intersection of two sets of possibilities: x and not-x, and y and not-y. Explain this using examples rather than these abstract terms. For example, if the children want to write 'like basketball' and 'like football' as the column headings, explain that they need to write 'like basketball' and 'don't like basketball' instead.

Answers for Pupil Book 3 page 77

1. a 4
   b Teresa
   c Mandla, Dan and Connie
   d Ahmed and Bali

2–3 Individual answers

Answers for Workbook 3 page 63

1

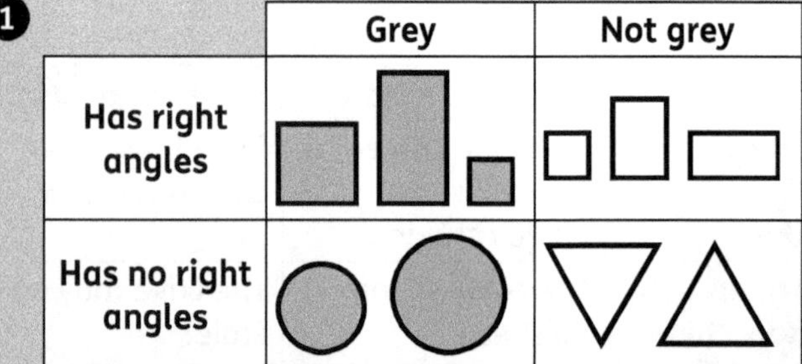

	Grey	Not grey
Has right angles		
Has no right angles		

2 a 11   **b** 5   **c** 2   **d** 1   **e** 3

3

	Even	Not even
Multiple of 5	10, 20, 30, 50	5, 15, 35
Not a multiple of 5	2, 4, 6, 8, 12, 26	1, 3, 7, 9

Answers for Workbook 3 page 64

1 Possible answer: In the fridge because we want to protect our food from mini-beasts such as ants, beetles and spiders. We would find lots of mini-beasts living under stones and logs and among flowers.

2–**3** Individual answers

End-of-unit check

To assess the children's understanding of the different ways to show data, ask questions such as:

- Show a pictogram. *What does this symbol represent?*
- *How could I show* (give a number) *ice creams/cars/ books using this symbol?*
- *How can we use a bar chart to find out which item is the most popular?* (Find the longest bar.)
- *How do we show numbers up to five with tally marks?* (I, II, III, IIII, ⅢⅠ)

Ask questions about the data shown in tables, bar charts, pictograms, Venn diagrams or Carroll diagrams. The children must be able to find the answers by looking at the table or diagram.

Mixed practice 2

You can use Mixed practice 2 on **Pupil Book 3 pages 78–79** to assess the children's confidence in the concepts from Units 7–12.

Answers for Mixed practice 2, Pupil Book 3 page 78–79

1 a Possible answers: $1 + 9 = 10$, $3 + 7 = 10$, $5 + 5 = 10$
b Possible answers: $10 + 90 = 100$, $30 + 70 = 100$, $50 + 50 = 100$

2 a \$464   **b** \$900

3 a 25
b Possible answer: Rounding down would give 24 packs, which is 240 cards and so not enough.

4 a 113   **b** 230   **c** 550
d 145   **e** 649   **f** 161

5 a \$0.50 or 50c   **b** \$0.10 or 10c
c \$0.75 or 75c   **d** \$1.80

6 a More than 1 kg   **b** About 1 kg
c Less than 1 kg

7 10

8 a $\frac{1}{2}$ kg or 500 g   **b** 1 kg   **c** 4 kg

9 $3 \times 6 = 18$, $6 \times 3 = 18$, $18 \times 6 = 3$, $18 \times 3 = 6$

10 a 3, 6, 9, 12, 15, 18, 21, 24, 27, 30
b 5, 10, 15, 20, 25, 30, 35, 40, 45, 50
c It does not end in 0 or 5.

11 He has worked out the area, not the perimeter.

12 a 8 cm²   **b** 16 cm²   **c** 13 cm²

13 9 cm²

14 a Junk   **b** 4   **c** 2

15 Venn diagram drawn as follows:
in 'Set A' circle only: 102, 104, 106, 108, 112, 114, 116, 118
in 'Set B' circle only: 90, 95, 105, 115, 125
in the overlapping section: 100, 110, 120

16 Individual answers. For example:

	Edible	Not edible
Red	strawberry	brick
Not food	banana	tree

UNIT 13 — 3D shapes

Learning objectives
- Identify, describe, sort, name and sketch 3D shapes by their properties.
- Recognise pictures, drawings and diagrams of 3D shapes.
- Draw 2D shapes and make 3D shapes using modelling materials.
- Recognise 3D shapes in different orientations and describe them.

Key words
3D shape   cube   cuboid   cone   cylinder   prism   pyramid   face   vertex (vertices)   edge   triangular prism   square-based pyramid   net

Unit introduction

Materials
3D shapes for demonstration (cube, cuboid, cylinder, triangular prism, square-based pyramid); coloured paper or card, scissors, magazines and other printed material (for the children to cut out pictures of 3D shapes used in a variety of ways)

Teaching guidance
The children should know the names of some *3D shapes* from their work at Level 2. However, they may have forgotten some and they may not know the correct terms for parts of the shapes.

Make sure you talk about 3D shapes using their correct names and that you use the correct terms (*face, vertex* and *edge*) when you refer to the parts of a shape. Using the correct terms and insisting that the children use them will allow the children to talk about 3D shapes in the correct way.

At Level 2, the children sorted shapes according to their properties. Now they will classify 3D shapes based on the number of faces they have and the shapes of the faces.

The children will begin to learn that prisms have two matching (congruent) end faces and that pyramids all have a base and triangular faces that meet at a vertex. It is important to show the children examples of the 3D shapes so they can see all the parts.

Children don't always see 2D drawings of 3D shapes as shapes. In one case where children were shown pictures of a cube built with straws and asked to build the cube, they built a replica of the 2D picture and not a 3D cube. Lots of practical activities will help the children understand the 2D diagrams of 3D shapes that they will encounter in mathematics.

- Divide the class into groups and give each group a set of 3D shapes (*cubes, cuboids, cones, cylinders, prisms, pyramids, spheres*). Get the children in the group to name the shapes.
- Working in pairs, one child describes the shapes of the faces of a 3D shape and the other child names the shape.
- The children could carry out a survey of the classroom or school to identify different 3D shapes and describe how they are being used.
- As a homework activity, ask the children to describe the shapes of some real objects at home.
- The children can make a scrapbook of pictures of 3D shapes from magazines and other publications. Ask them to bring magazines into school prior to this activity.
- Lead a whole-class activity. Show the children different shapes and ask them to identify the shapes of the faces and the number of them. Use a cube and a cylinder to demonstrate the difference between flat surfaces and curved surfaces.

Support
Display six 3D objects, labelled A–F. Display the correct names of the shapes in a mixed order and ask the children to match the shapes with their names.

Identify 3D shapes

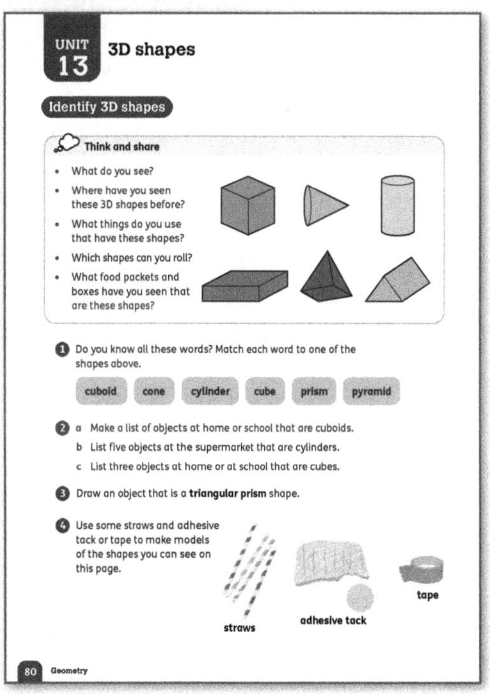

Materials
Feely bags (page 23) containing small 3D shapes; sets of mixed 3D shapes; construction materials (straws or toothpicks and adhesive tack or modelling clay) or construction sets

Warm-up

- Give each group of children a feely bag of small 3D shapes. The children in each group take turns to put their hands in the bag and feel for a shape without removing it. They should describe the shape to their group and the group should try to guess what 3D shape it is. The child can then take out the shape for the group to see.
- Spend time classifying 3D shapes in different ways. Give each group a set of mixed 3D shapes and ask them to sort them into two groups. They should say what criteria they used to group the shapes. Then ask them to sort the shapes again, this time making three (or four) groups. Again, give them time to explain the criteria they used. This activity allows the children to study the shapes closely and to see that there are different ways of classifying the shapes using properties and characteristics.

Focus

- <u>Think and share:</u> Discuss the 3D shapes shown at the top of **Pupil Book 3 page 80** and ask the children the questions about these shapes. Revise the names of these shapes.
- Read through the questions with the class, then let the children complete the questions to check their understanding.
- In question 1, the children could copy the drawings of the shapes and label them in their notebooks.
- If necessary for question 2, you could provide a range of objects or pictures of objects that are cuboids, cylinders and cubes to help the children.
- In question 3, you may need to show the children how to draw a *triangular prism*: draw two identical triangles and then join the corresponding corners with straight lines.
- Once the children are confident with the names and properties of 3D shapes, give them construction materials and let them spend some time building their own models of 3D shapes as detailed in question 4. Demonstrate how to use drinking straws and lumps of adhesive tack to build frameworks of shapes. The children could also use construction sets if you have them.
- As noted in the Support section below, you can also wait until later in the unit for the construction activity in question 4 if you prefer.

Challenge

The children can use construction kits or other materials to make compound shapes (shapes made by combining two or more different 3D shapes), for example a square-based pyramid on a cube.

Support

Some children may need help to build their models. If you wish, do the construction activity in question 4 a little later in the unit, when the children have had more time to develop their understanding of faces, edges and *vertices*.

Some children may be confused by the terminology of 2D and 3D shapes. Regular use will reinforce the correct terms. Play games in which the children describe shapes. Read out descriptions of 2D and 3D shapes and ask the children to identify them.

Interesting mistakes

Some children may confuse cubes and cuboids. Remind them that a cube is a special type of cuboid in which all of the sides are exactly the same length and all the faces are squares. Likewise, a square is a special type of rectangle in which all of the sides are the same length.

The children are likely to confuse pyramids and prisms. Invite them to compare what is similar and what is different about these shapes, and ask them to identify how we can work out whether a shape is a pyramid or a prism. Once the children understand that a prism has two parallel ends, they can explore some other types of prisms.

Answers for Pupil Book 3 page 80

<u>Think and share:</u> six 3D shapes: cube, cone, cylinder, cuboid, pyramid, prism
Individual answers. For example:
- objects in my home, shapes on buildings, in the classroom
- my Rubik's cube, my water bottle, the Pyramid of Cheops

I can roll the cone and the cylinder.
Individual answers. For example: cereal boxes, jigsaw box, the laundry basket, a tube of sweets

1 top row: cube, cone, cylinder
bottom row: cuboid, pyramid, prism
2–**3** Individual answers
4 Practical work

Faces, edges and vertices

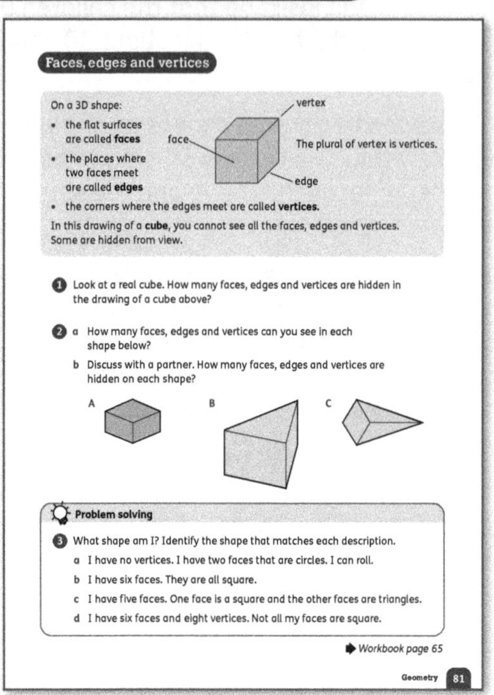

Materials

Sets of 3D shapes; sticky notes or small cards

Warm-up

- Have the children investigate the shapes of the faces of different 3D shapes.
- They can use sticky notes or small cards to stick the name of the 2D shape onto each face of the 3D shape. They can then remove these labels and count and/or draw them.
- The children can also trace around 3D shapes to see the shapes of the faces.
- If you have plastic shapes that fit together to build 3D shapes, you can prepare activities in which the children construct and deconstruct the shapes to find out more about the faces.

Focus

- Use the information at the top of **Pupil Book 3 page 81** to revise the terms *face*, *vertex* (and *vertices*) and *edge*.
- Show the children a cube and discuss the fact that we cannot see all the parts at once, so some parts are not visible in a drawing of the cube.
- Let the children work through questions 1–3 in pairs or groups and allow them to use their models from the previous lesson or other 3D shapes to check their answers.
- <u>Problem solving:</u> In question 3, the children use clues to identify mystery shapes.

Follow-up

Workbook 3 page 65 assesses the children's ability to name 3D shapes and identify their properties, as well as their ability to use a table correctly to summarise information.

Challenge

Challenge the children to try to use 3D shapes to make other 3D shapes. For example, say:

- *Put together two 3D shapes to make a shape with eight faces and six vertices.*
- *Put together two triangular prisms and describe the new 3D shapes you have made.*
- *What other 3D shapes can you make if you put the prisms together in other ways?*

Support

The questions on visible and hidden faces, edges and vertices are very challenging. For the children who find this too difficult, focus on counting the faces, edges and vertices of known 3D shapes. Then hold up 3D models of the shapes and ask: *How many faces can you see when I hold it like this?* Then, after turning the 3D shape to a new position, ask: *Now how many faces can you see?* You can do the same with vertices and edges.

Interesting mistakes

Some children will confuse the number of sides of a face with the number of faces. For example, they may say a cube has four faces (because a square has four sides) or a pyramid has three faces (because a triangle has three sides). Provide plenty of practice in identifying and counting the faces of 3D shapes.

Answers for Pupil Book 3 page 81

1 3 faces, 3 edges, 1 vertex

2 a A 3 faces, 9 edges, 7 vertices
    B 3 faces, 8 edges, 7 vertices
    C 2 faces, 5 edges, 4 vertices
  b A 3 faces, 3 edges, 1 vertex
    B 2 faces, 1 edge, 0 vertices
    C 3 faces, 3 edges, 1 vertex

3 a sphere b cube c pyramid d cuboid

Answers for Workbook 3 page 65

1 cube: 6 faces, 12 edges, 8 vertices
cuboid: 6 faces, 12 edges, 8 vertices
triangular prism: 5 faces, 9 edges, 6 vertices
cylinder: 2 faces (and 1 curved surface), 2 edges, 0 vertices
pyramid: 5 faces, 8 edges, 5 vertices

2 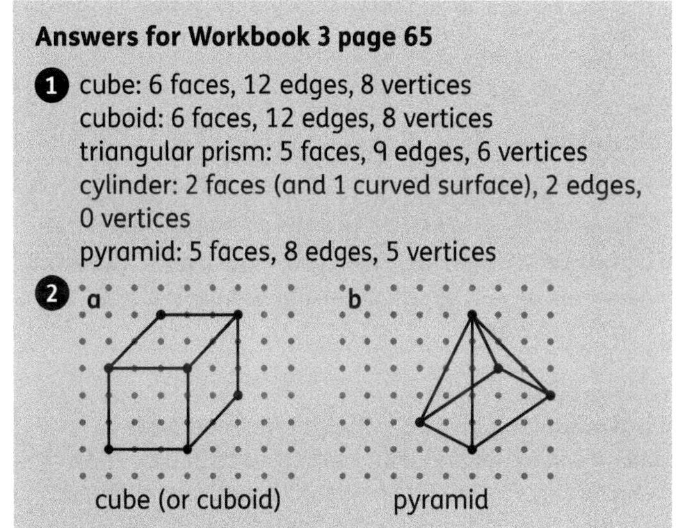

cube (or cuboid)　　　　pyramid

Explore 3D shapes

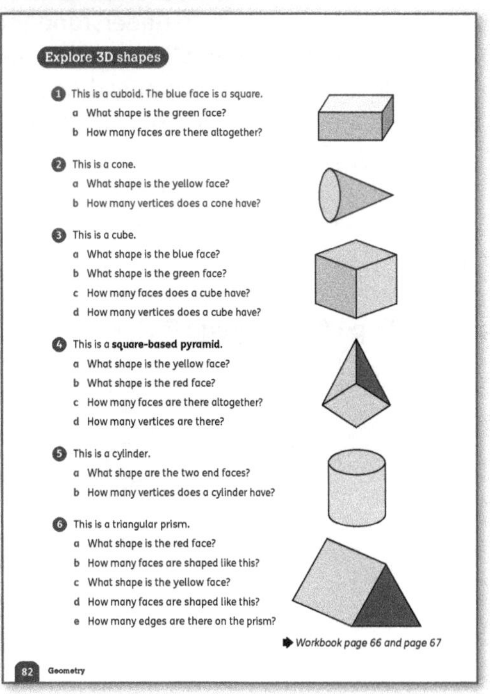

Materials

Sets of 3D shapes

Warm-up

- Prepare sets of 3D shapes in which one shape is the 'odd one out' and ask the children to decide which one it is. For example, include a cube in a set of cuboids, a prism in a set of pyramids, or a cylinder in a set of cuboids.
- Introduce the term *square-based pyramid* and point out the square face at the bottom.

Focus

- The children could use Carroll diagrams to sort a set of 3D shapes using different characteristics. For example, they can use this one to sort cubes and cuboids from other 3D shapes:

	Has six faces	Does not have six faces
Faces are all square		
Faces are not all square		

- Use **Pupil Book 3 page 82** together with **Workbook 3 page 66** to consolidate the work on 3D shapes and parts of 3D shapes. Again, you may need to provide models of 3D shapes so the children can check their answers.

Follow-up

On **Workbook 3 page 67**, the children could base their drawings of buildings and other structures on real buildings in the local area. Alternatively, they can use their imaginations to draw buildings and other structures that include familiar 3D shapes.

Support

The children may need support to understand that the questions on **Pupil Book 3 page 82** are asking about all the faces, edges or vertices, not only those that are visible. Note also that the drawing of a 3D shape distorts the shape visually, so the circle at the top of the cylinder looks like an oval. If this confuses the children, keep showing them concrete models of the 3D shapes as you ask the questions.

Answers for Pupil Book 3 page 82

1 a rectangle　　b 6

2 a circle　　b 1

3 a square　　b square　　c 6　　d 8

4 a square　　b triangle　　c 5　　d 5

5 a circle　　b 0

6 a triangle　　b 2　　c rectangle
　　d 3　　e 9

Answers for Workbook 3 page 66

1 cylinder, sphere, pyramid
prism, cone, cuboid

2 Individual answers

3 Cylinder: 9　　Sphere: 4　　Pyramid: 1
Triangular prism: 5　　Cone: 8　　Cuboid: 7

Answers for Workbook 3 page 67

1 a the roof of one of the houses
　b cuboids and triangular prisms (and one with cuboid and pyramid)
　c cylinders　　d pyramid
　e a water tank, bins/barrels outside a house, towers

2 Individual answers

3D shapes in different positions

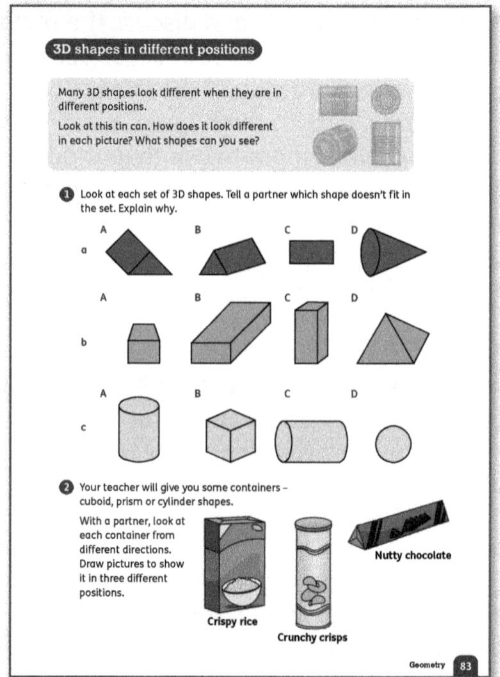

Materials

Marker pen or eraser; 3D shapes; empty containers that are cuboid, prism or cylinder shapes – boxes, tins, tubes and other food packets

Warm-up

- Discuss the tin can in the pictures on **Pupil Book 3 page 83**. Ask questions such as:
  - What shape is the tin can? (cylinder)
  - Do you recognise this shape?
  - What else have you seen that is this shape?
  - Why does it look like a circle in one of the pictures? (We are looking at it from one end.)
- Ask the children to use an object such as a marker pen or an eraser. The children move their object into the same positions as the tin can.

Focus

- Work through question 1 with the class, discussing each set of shapes and which one doesn't fit. This is a challenging exercise. Let the children discuss with a partner, and then with the whole class, what they see in each picture.
- In question 2 hand out the containers you have brought. Let the children work in pairs to draw a container in three different positions.

Challenge

Some children may enjoy the challenge of trying to draw 3D shapes in a variety of positions without referring to a concrete shape. For example, you could challenge the children to draw a pyramid from above, from below and from the side.

Support

Some children may have difficulty drawing 3D shapes in different positions. Their drawings do not have to be precise. Encourage them to describe what they see, for example:

- *How does this 3D shape look from this side?*
- *Does it look long or short?*
- *Does it look round?*
- *Can you see any corners?*
- *Which faces can you see?*
- *Start drawing the largest face you see. Now add more.*

Continue with similar questions.

> **Answers for Pupil Book 3 page 83**
>
> **1** Possible answers:
>   **a** D because it is not a prism
>   **b** D because it is not a cuboid
>   **c** B because it is not a cylinder
>
> **2** Individual answers

Nets of cubes

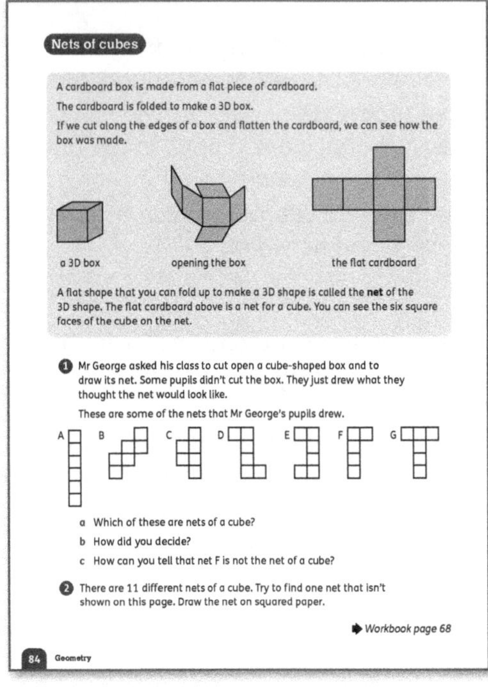

Materials

Boxes to take apart, scissors; thin card or stiff paper, glue, sticky tape; squared paper

Warm-up

- Use a cube-shaped box and show the children how to cut it apart and flatten it to make a *net*.
- If you cannot find a cube-shaped box, make your own using an enlarged version of the net on **Workbook 3 page 68**.
- Let all the children trace or copy the net in the **Workbook** to make their own cube.

Focus

Let the children work in pairs to complete questions 1 and 2 on **Pupil Book 3 page 84**. If necessary, provide squared paper and allow the children to cut out and fold up the nets shown to see if they can be folded to make cubes.

Support

The net-building activity will take some time. Some children may find it challenging, so you may need to assist them.

> **Answers for Pupil Book 3 page 84**
>
> **1** **a** B, D, G
>   **b** Possible answer: Make the nets and fold to make a cube.
>   **c** There are only 5 squares and a cube has 6 faces.
>
> **2** The children should find one or more of these nets:
>
> 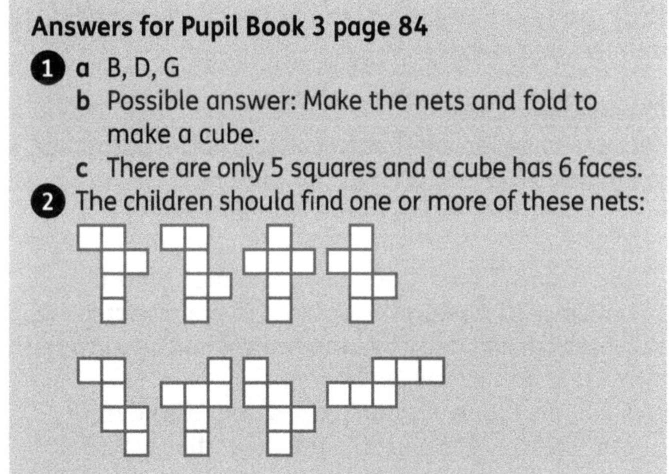

> **Answers for Workbook 3 page 68**
>
> **1** Practical work making a cube

End-of-unit check

Assess the children's understanding of 3D shapes by asking questions such as:

- *What shape is a box of cereal?* (cuboid)
- *How many faces does a triangular prism have?* (5)
- *Can you name a 3D shape that has all flat faces?* (cube, cuboid, pyramid, any prism (e.g. triangular prism))
- *Can you name a 3D shape that has flat faces and curved surfaces?* (cylinder, cone)
- Show the children a set of shapes. *How could you group these 3D shapes into sets? How do you decide whether a shape fits into a set?*
- *How many faces/edges/vertices does this 3D shape have?*
- *Can you describe this 3D shape to me?*
- *Can I make a cube with this net? Why or why not?*

Learning objectives
- Interpret and create descriptions of position, direction and movement including reference to cardinal points.

Key words
clockwise   anti-clockwise   quarter turn
half turn   three-quarter turn   full turn   position
above   below   under   next to   between
direction   up   down   left   right   forwards
backwards   cardinal points   north   south
west   east   grid   block   row   column

Unit introduction

Materials
Clock; cards or labels; a chessboard or checkers board, counters and stickers

Teaching guidance
Here are some ideas for practical activities to introduce the terminology and concepts of this unit:
- Demonstrate *clockwise* and *anti-clockwise* movements. Show the children the direction in which the hands of the clock move. Do some practical activities that allow the children to practise creating clockwise and anti-clockwise movements – walking, making arm circles, and so on.
- Label the four walls of the classroom with the numbers 12, 3, 6 and 9 (in the order they would appear on a clock face). Demonstrate what we mean by a *quarter turn, half turn, three-quarter turn* and *full turn*. Let the children take turns to direct each other between the desks to a place they have marked. They should give the direction and the distance, for example: 'Make a quarter turn clockwise and walk past three desks.'
- If possible, label shelves or lockers in the classroom with letters and numbers (use **Pupil Book 3 page 87** as a reference). Have the children use the letters and numbers to identify various lockers or *positions* on the shelves.
- Show the class a game board with squares (chess or checkers) and place a counter on one square of the board. Let the children suggest how they could identify the position of the square that the counter is on. They could then use stickers to label the rows and columns with numbers and letters.

Position and direction

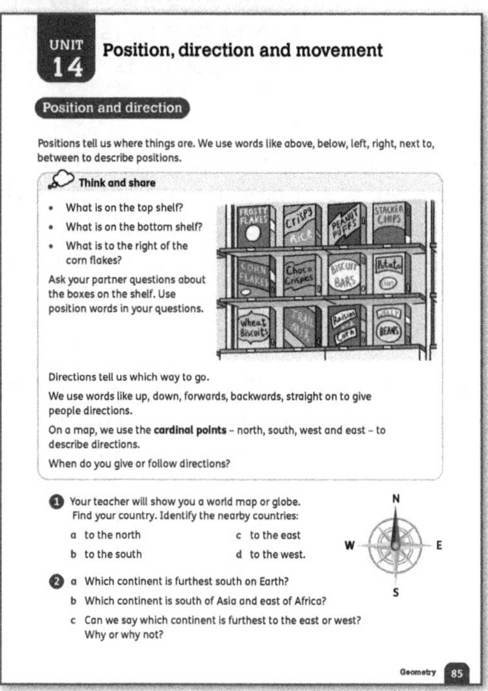

Materials
A world map or globe; a compass

Warm-up
- <u>Think and share:</u> Turn to **Pupil Book 3 page 85**. Ask the children what they see in the picture of the supermarket shelves. Ask the questions, as well as more questions using *direction* words. For example, ask:
  - *What is on the top shelf?*
  - *What is on the bottom shelf?*
  - *What is to the left of the* (named product)*?*
  - *What is to the right of the* (named product)*?*
  - *What is between the* (named product) *and the* (named product)*?*
  - *What is furthest to the right/left?*
  - *What is above the* (named product), *but below the* (named product)*?*
- The children should work in pairs to ask questions about the positions of items on the shelves. Encourage them to use words such as *above, below, under, next to* and *between*.

Focus
- Show the world map or globe and invite the class to identify your country and your neighbouring countries.
- Introduce or revise the *cardinal points* (the four main directions on a compass): *north, east, west* and *south*.
- Let the children identify neighbouring countries in each direction on the map or globe.

- Discuss the continents and point out the southern and northern hemispheres on the globe or map.
- Discuss questions 1 and 2 in **Pupil Book 3 page 85** with the class.

Challenge
You could extend the 'Think and share' activity to include directions. Ask the children to give their partner directions to items on the shelves, for example: 'Start at the crispy rice. Move down two shelves, then move left. What do you find?'

For the children who are interested, you can challenge them to locate given countries or cities based on direction 'clues'. Ask, for example: *Which country is east of Malaysia? Which country is north of India?*

Support
Make sure the children have a firm grasp of the meanings of *up, down, left* and *right* before you move on to the cardinal points.

Interesting mistakes
Some children may mistakenly want to identify the cardinal points as fixed positions or places. It is important for them to realise that these are not fixed positions, but directions that describe relationships between places. On a map, north, west, south and east correspond to up left, down and right.

> ### Answers for Pupil Book 3 page 85
> <u>Think and share:</u> Frosty Flakes, Crispy Rice, Peanut Puffs and Stacker Chips
> Wheat Biscuits, Trail Mix, Raisins and Corn,
> Jelly Beans
> Choco Crispies
> Possible questions: What is to the left of the crispy rice? What is below the frosty flakes? What is between the choco crispies and the potato chips? Individual answers. For example: I follow directions when I want to find a classroom in the school. I give my friend directions to my home.
> **1** Individual answers, depending on the country
> **2** a Antarctica   b Australia
>   c No, because the Earth is a sphere, so it is always possible to go further east or west.

Clockwise and anti-clockwise

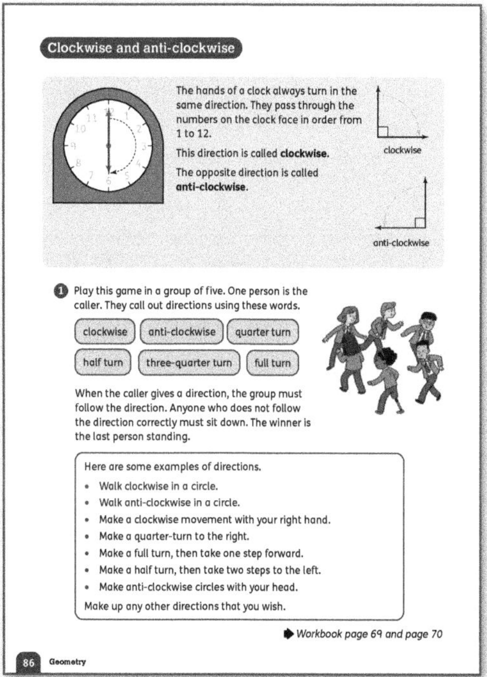

Materials
Clock with moveable hands

Warm-up
Use a clock with moveable hands to show clockwise and anti-clockwise turns. When you make each turn, ask the children to say whether you have turned the hands clockwise or anti-clockwise. You can also revise full, three-quarter, half and quarter turns.

Focus
Let the children play the game described in question 1 of **Pupil Book 3 page 86** in groups of five. You may wish to take the class into the hall or the playground to give them more space.

Follow-up
- Once the children understand clockwise and anti-clockwise directions, let them work alone or in pairs to complete **Workbook 3 page 69** and **page 70**.
- Discuss the children's drawings of circles in question 3 on **Workbook 3 page 70**: *Which motion did you find easier or more comfortable: clockwise or anti-clockwise? Can you explain why?*

Support
- Some children may take a while to understand the concepts of clockwise and anti-clockwise.
- Over a week or two, take a few minutes at the start of each lesson to help the children notice the direction that the hands are moving around the classroom clock face (if you have a classroom clock).

> ### Answers for Pupil Book 3 page 86
> **1** Practical activity

Answers for Workbook 3 page 69

1

Starting position	Turn	End position
Facing the tree	$\frac{1}{2}$ turn clockwise	Facing the swing
	$\frac{1}{2}$ turn anti-clockwise	Facing the swing
Facing the swing	$\frac{1}{4}$ turn anti-clockwise	Facing the house
	$\frac{1}{4}$ turn clockwise	Facing the pond
Facing the pond	$\frac{3}{4}$ turn anti-clockwise	Facing the tree
	$\frac{3}{4}$ turn clockwise	Facing the swing
Facing the house	$\frac{1}{2}$ turn clockwise	Facing the pond
	$\frac{3}{4}$ turn clockwise	Facing the tree
Facing the tree	$\frac{1}{4}$ turn anti-clockwise	Facing the pond
	$\frac{3}{4}$ turn clockwise	Facing the pond

2 Individual answers

Answers for Workbook 3 page 70

1   a   1, 4, 5   b   2, 3

2

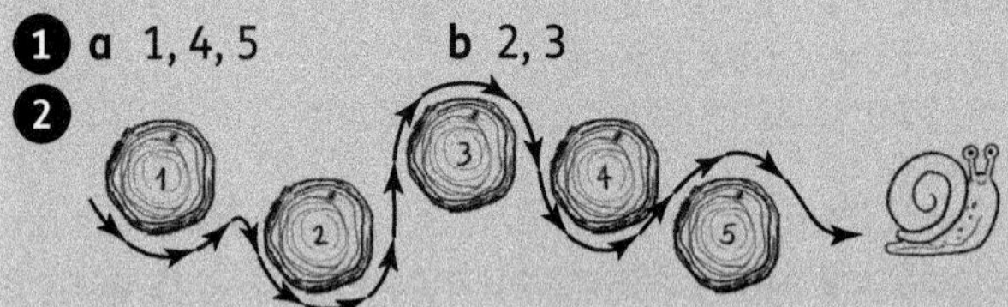

3 Individual answers. For example: 'I write with my right hand, so I find it easier to draw in a clockwise direction, starting from the left.'
'I write with my left hand, so I find it easier to draw in an anti-clockwise direction, starting at the right.'

Position on a grid

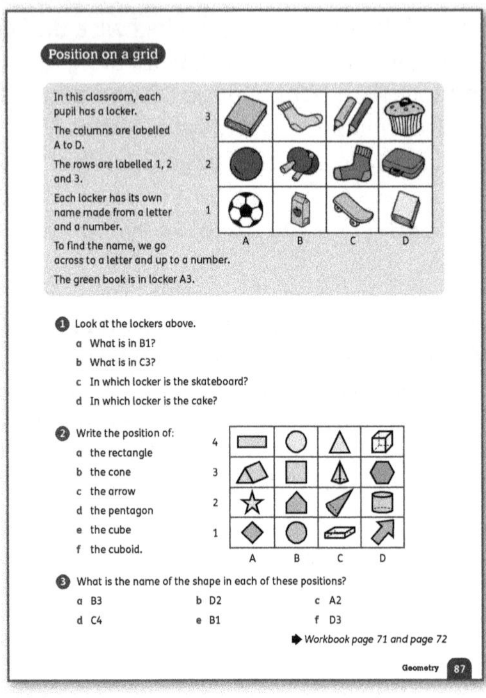

Materials
A large grid with columns and rows labelled (letters on the horizontal axis and numbers on the vertical axis); shapes or objects to place on the grid

Warm-up
On your prepared *grid*, place 2D or 3D shapes in the *blocks* and ask questions such as:
* *Where is the red triangle?*
* *In which block is the blue square?*
* *What is in C3?*

Focus
* Use **Pupil Book 3 page 87** to teach the children how to use grid coordinates made up of a letter and a number.
* Work through questions 1–3 with the class.
* Revise the terms *row* and *column*, then demonstrate how we name coordinate blocks using first the letter then the number.

Follow-up
Once the children are able to locate the position and give the coordinates of an item, let them work in pairs to complete **Workbook 3 page 71** and **page 72**. Let each pair check another pair's completed grid on **Workbook 3 page 72**.

Challenge
Any children who finish the activities quickly and easily might enjoy making up questions for a partner about the grid on **Workbook 3 page 72**. They could:
* Write directions from one shape to another, using 'up', 'left', 'right' and 'down'.
* Give clues about the position of a shape by describing the shapes around it (for example: 'I have a triangle to my left and a hexagon below me.').

Answers for Pupil Book 3 page 87

1   a   juice   b   pencils
    c   C1   d   D3

2   a   A4   b   C2   c   D1
    d   B2   e   D4   f   C1

3   a   square   b   cylinder   c   star
    d   triangle   e   circle   f   hexagon

Answers for Workbook 3 page 71

1

Back

4		Dahlia	Nina		Zayed
3			Bess		
2	Jo	Rick	Thandi	Cara	Luke
1	Ella		Malala	Amani	Sipho
	A	B	C	D	E

Front

2   a   E4   b   E1   c   D2
    d   A2   e   C2

3   a   E3   b   A3   c   B1

Answers for Workbook 3 page 72

Answers for Workbook 3 page 72

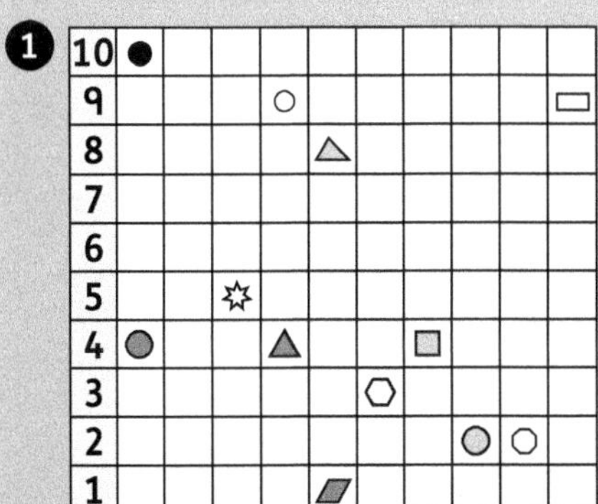

2 a Possible answers: 5 blocks left (or west) and
7 blocks up (or north)
or: start facing up (or north), move 7 blocks
forwards, quarter turn anticlockwise, move
5 blocks forwards

b Possible answers: 4 blocks down (or south)
and 7 blocks left (or west)
or: start facing down (or south), move 4 blocks
forwards, quarter turn clockwise, move 7
blocks forwards

Move a shape

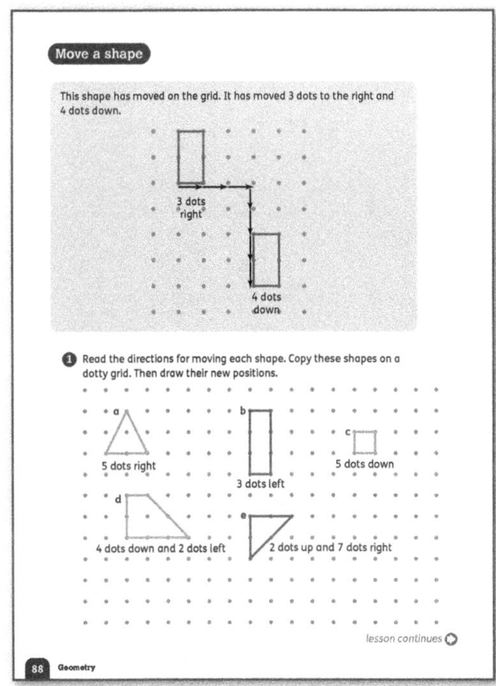

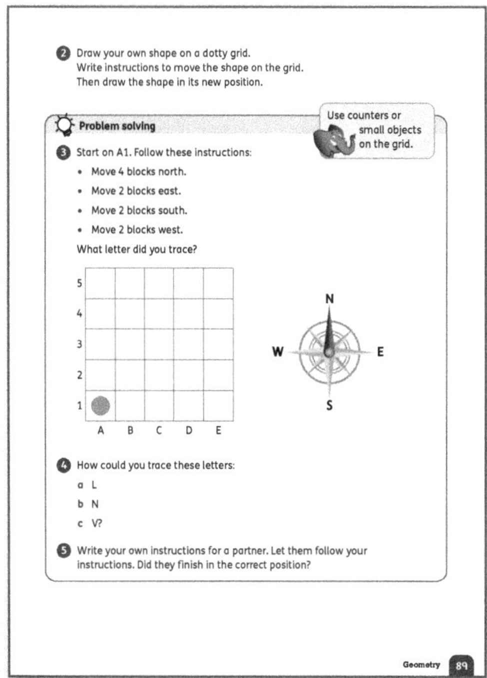

Materials
Dotted or squared paper, rulers; counters

Warm-up
- **Pupil Book 3 page 88** and **page 89** introduce slides
(translations) on a grid in an informal way. The children
begin following and giving instructions to change the
position of shapes on a grid. This will prepare them for
later work on transformations (slides, flips and turns, or
translations, reflections and rotations).
- Discuss the picture and the information at the top of
Pupil Book 3 page 88 with the class. Ask questions
such as:
  - How has the position of the shape changed?
  - Which way did the shape move?
- Encourage the children to notice that there are
different ways to get from one position to another,
by asking:
  - What other instructions could you give to move
the shape to that position?

Focus
- Hand out dotted or squared paper for the questions
on **Pupil Book 3 page 88** and **page 89**.
- In question 1, the children copy the shapes they see on
the grid, and then draw each shape in its new position.
- In question 2, the children draw their own shape on
dotted grid paper and write instructions for how they
want the shape to move. They draw the new position
of the shape.
- <u>Problem solving</u>: Questions 3–5 require the children
to trace letters using a grid. In questions 3 and 4 the
children solve the problem – they can place a counter
on the grid to do this. In question 5 they write their
own instructions for a partner to follow.

Challenge
The children can write an alternative set of instructions to
move each shape from its first position to its final position.

1

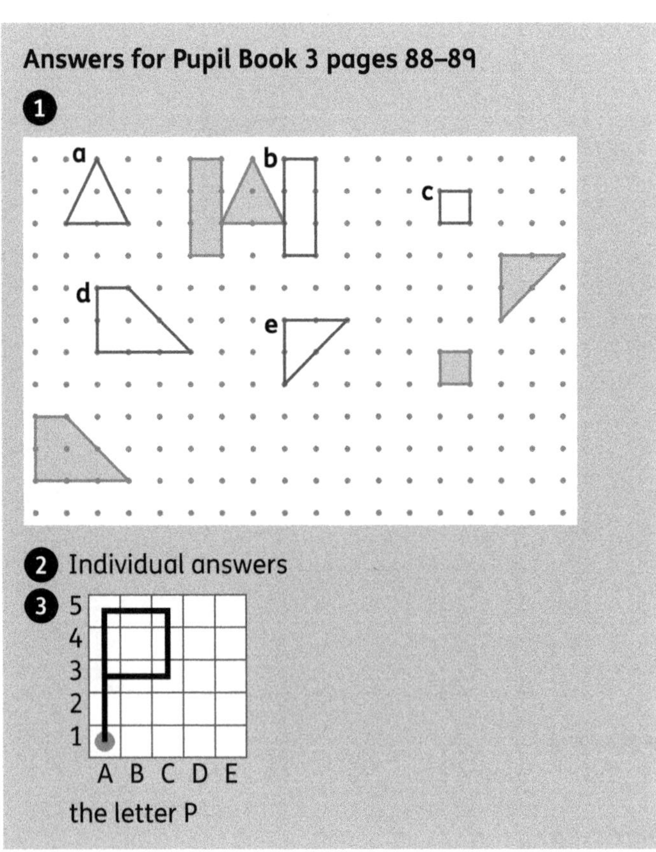

2 Individual answers

3
the letter P

4 Possible answers:
  a Start on A5. Move 4 blocks south. Move 2 blocks east.
  b Start on A1. Move 4 blocks north. Move 3 blocks east and 4 blocks south. Move 4 blocks north.
  c Start on A5. Move 2 blocks east and 4 blocks south. Move 2 blocks east and 4 blocks north.

5 Individual answers

End-of-unit check

Ask questions using the vocabulary of direction, such as:
- *Can you point your arm* (or an object, if more appropriate) *up/down/left/right?*
- *Now can you make a quarter/half/three-quarter turn clockwise/anti-clockwise?*

Give the children a coordinate grid like the ones on **Pupil Book 3 page 87**, with objects or shapes drawn in some of the blocks. Ask questions about it, such as:
- *Where is the …?*
- *What is in block A4?*
- *Which block is to the east of block C3?*
- *In which block is the* (named object or shape)*?*

UNIT 15 Fractions

Learning objectives
- Interpret and write proper fractions.
- Recognise, find and write fractions of a discrete set of objects: unit fractions and non-unit fractions with small denominators.
- Recognise and use fractions as numbers and fractions of a set of objects: unit fractions and non-unit fractions with small denominators.
- Count up and down in tenths.
- Recognise that tenths arise from dividing an object into 10 equal parts and in dividing 1-digit numbers or quantities by 10.
- Understand and explain that fractions are several equal parts of an object or shape and all the parts together equal one whole.
- Understand and explain that fractions can describe equal parts of a quantity or set of objects.
- Recognise and show, using diagrams, equivalent fractions with small denominators.
- Compare and order unit fractions, and fractions with the same denominator.
- Add and subtract fractions with the same denominator (within one whole).
- Solve problems involving fractions.

Key words
fraction half third quarter/fourth fifth sixth tenth numerator denominator equivalent fractions greater than less than

Unit introduction

Materials
Large coloured shapes divided into different fractions; sheets of paper; items to cut in half (modelling clay, bread, string, etc.); simple jigsaws (made by gluing a picture onto card and cutting into 2, 3, 4, 5, 6 or 10 equal pieces); counters

Teaching guidance

Use any of these activities to introduce the work on *fractions*:

- Discuss how we can divide an object or shape into parts. Use a simple shape, such as a circle or a square. Demonstrate, by drawing a line, that the shape can be divided into two parts in lots of different ways. Some lines create two equal parts and others create two unequal parts. Repeat the process, this time drawing two lines to show how the shape can be divided into four equal parts.
- Revise or introduce the words we use to describe the parts formed when an object is divided into parts:
  - two parts – halves
  - three parts – thirds
  - four parts – fourths or quarters
  - five parts – fifths
  - ten parts – tenths.
- Ask the children how they could divide a sheet of paper into two or four equal parts. Then ask them where they would need to draw a line or lines to divide their desk, the board, the classroom, etc. into two and four equal parts. You could give them suitable items to cut in half, such as a piece of modelling clay, a slice of bread or a length of string.
- Give the children a simple jigsaw made from a picture cut into equal pieces. Making the jigsaw will allow the children to experience the number of parts in the whole in a practical way.
- Divide a square into four equal parts. Demonstrate that one part is equal to one-quarter (or fourth), two parts are equal to two-quarters and that three parts are equal to three-quarters. Ask them: *Which of these fractions is the same as one-half?* (2 parts) Repeat this for other fractions.
- Use a simple shape, such as a circle or square, to demonstrate dividing an object into thirds (three equal parts). Then show that we can divide each third into two parts to make six parts, called *sixths*. Ask: *How many sixths are equal to one-third/ two-thirds/three-thirds?* (2/4/6)
- The children could carry out a survey of uses of fractions in everyday life.

Parts of a whole

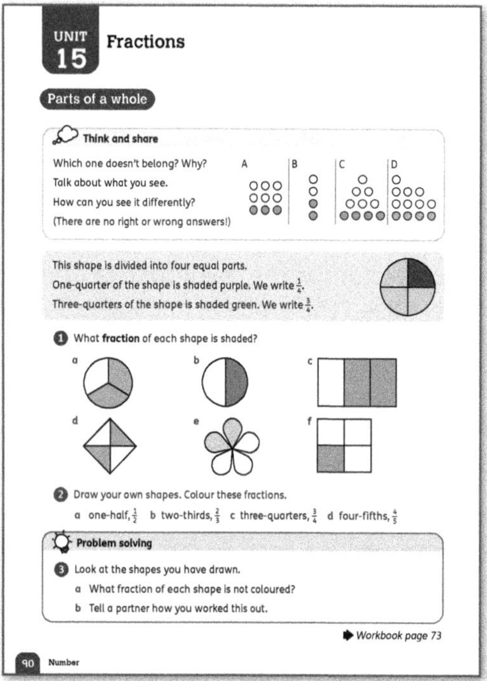

Materials

Paper plates or circles cut from paper or card

Warm-up

- <u>Think and share:</u> The children work in small groups to discuss the four diagrams at the top of **Pupil Book 3 page 90**. As well as the questions in the book, ask:
  - What do you notice?
  - What is the same? What is different?
  - What do you wonder?
- There is no single correct 'odd one out'. The answer depends on how we look at the diagrams. The most obvious odd one out is D, which has dots of a different colour. Here are some other possible answers and reasons:
  - D is the only shape that is not in a regular shape.
  - C is the only triangle.
  - A is the only shape with three rows instead of four.
- Gradually, encourage the children to start thinking about fractions. Ask: *How many of the dots in each group are shaded? What fraction is that?*
  - In diagram A, 3 out of 9 dots or $\frac{3}{9}$ or $\frac{1}{3}$ are shaded. The whole shape is made up of 3 rows, each with 3 dots.
  - In diagram B, 2 of the 4 dots or $\frac{2}{4}$ or $\frac{1}{2}$ are shaded.
  - Diagram C can be seen as $\frac{4}{10}$ or $\frac{1}{5}$.
  - Diagram D can be seen as $\frac{4}{12}$ or $\frac{1}{3}$.
- So, diagram C is the only one that does not show 1 of an equal number of parts (a unit fraction).
- It is not important if the class does not reach this point. It can be useful to leave them wondering about the question and come back to it later after the lesson on equivalent fractions **(Pupil Book 3 page 96).**

Focus

- In question 1, the children identify the fraction of each shape that is shaded. Keep emphasising that when we write a fraction, the number at the bottom is the total number of equal parts. The top number is the shaded part.
- In question 2, the children draw their own shapes and colour the fractions given.
- <u>Problem solving</u>: In question 3 children should be able to work out visually the fraction that is not coloured. Encourage the children to tell their partners how they solved the problem and invite some children to share their ideas with the class.

Follow-up

Use **Workbook 3 page 73** to consolidate this introductory work on fractions.

Challenge

Some children may like to draw interesting diagrams to show more unusual fractions such as $\frac{3}{11}$ or $\frac{4}{7}$. Remind them that all the parts must be equal in size.

Support

Cut paper plates or circles into fractions (quarters, halves, thirds, fifths) and let the children use the pieces to make wholes. Ask questions such as:

- *How many parts are there in this whole plate? Let's count them.*
- *What do we call one of these parts? (For example, point to one-quarter.)*
- *What do we call two of these parts?*
- *How many parts are there altogether?*

Interesting mistakes

Initially the children may be confused about the relationship between the numbers in a fraction. They may say, for example: '$\frac{1}{3}$ is 3 ones' or '$\frac{1}{3}$ is 1 three'. Encourage them to explain and understand their mistakes. When you are working with one-third, two-thirds, etc., refer back to the idea that this is a part of the whole. Remind the children that the whole is 1, and the fraction is less than 1. It is a part of 1, and so on.

Answers for Pupil Book 3 page 90

<u>Think and share:</u> Possible answers include: D has dots of a different colour. / D is the only shape that is not in a regular shape. / C is the only triangle. / A is the only shape with three rows. / C is the only diagram that does not show a unit fraction (1 of an equal number of parts). It shows $\frac{2}{5}$ of the whole.

Individual answers. For example: You can see it differently by moving dots to make different arrays.

1 a $\frac{2}{3}$ b $\frac{1}{2}$ c $\frac{2}{3}$

   d $\frac{2}{4}$ or $\frac{1}{2}$ e $\frac{2}{5}$ f $\frac{1}{4}$

2 Individual answers

3 a a $\frac{1}{3}$, b $\frac{1}{2}$, c $\frac{1}{3}$, d $\frac{1}{2}$, e $\frac{3}{5}$, f $\frac{3}{4}$

   b Possible answer: The number of unshaded sections at the top and the total of number of sections at the bottom.

Answers for Workbook 3 page 73

1 a 2 sections shaded   b 3 sections shaded

   c 2 sections shaded

2 one-quarter $\frac{1}{4}$ 4   one-half $\frac{1}{2}$ 2

   two-thirds $\frac{2}{3}$ 3   three-fifths $\frac{3}{5}$ 5

   five-sixths $\frac{5}{6}$ 6   four-tenths $\frac{4}{10}$ 10

3 a 5 green beads, 5 yellow beads

   b 3 green beads, 3 yellow beads, 6 purple beads

4 a 6 circles shaded   b 5 triangles shaded

More fractions

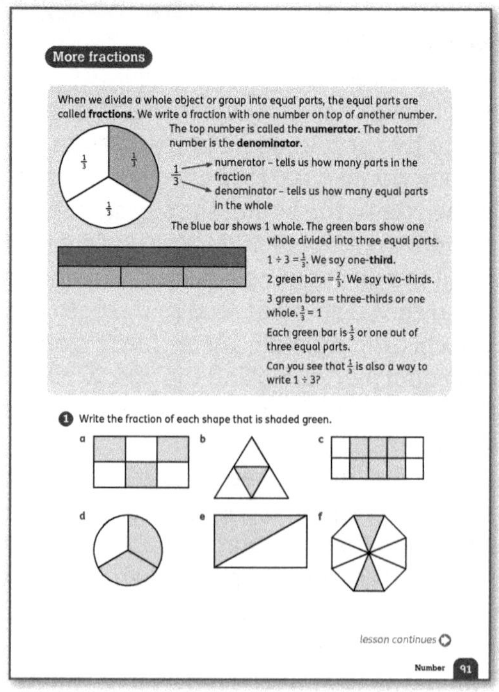

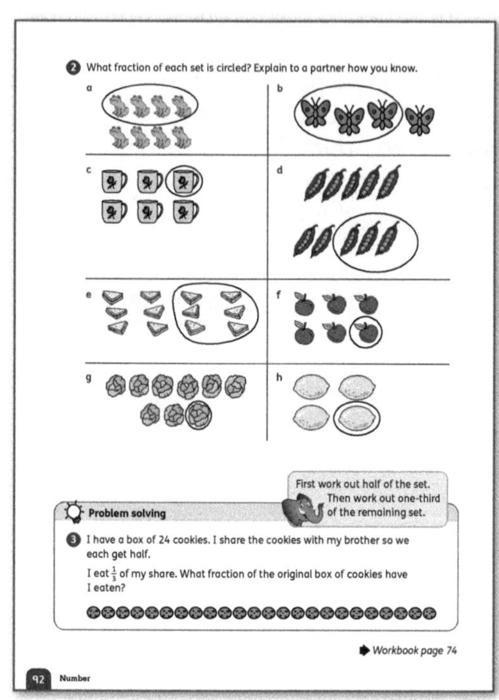

Materials

Number rods or strips of card about 20 cm long, marked as a whole, halves, thirds, quarters and tenths; fraction wall (you can search for this online); paper plates or circles cut from paper or card; sticky notes

Warm-up

* Hold up one of the strips of card. Tell the class: *This is one whole strip.*
* Hold up another strip. Say: *Is this the same as the other strip? … Yes, this is also one whole. Now, I'm going to fold it into 2 equal parts.* Fold the strip in half. *I have divided 1 into 2 equal parts. We can say 1 divided by 2 equals …*
* *How much is each part? Each part is half of the whole.*
* Write the division sentence: $1 \div 2 = \frac{1}{2}$. Say: *1 divided into 2 equal parts is equal to one half.*
* Repeat this with other strips of card, folding them into thirds, quarters, fifths, tenths, and so on.
* Make sure you relate the fractions to the division of the whole into parts.
* For fractions other than $\frac{1}{2}$, you can count through the fractions up to 1 (for example, *one-quarter, two-quarters, three-quarters, one whole*).
* The children may start noticing the equivalence of certain fractions to a half (two-quarters, five-tenths, and so on).

Focus

* Introduce the terms *numerator* and *denominator* to identify the top and bottom numbers in the fractions.
* You may also want to introduce the term *unit fraction* for fractions that have a 1 as the numerator (in other words, fractions that represent one of a number of equal parts).
* Not all fractions are unit fractions, and the children will need to distinguish between the denominator (which indicates the total number of equal parts) and the numerator (which indicates how many of those parts we are working with).
* Work through the example at the top of **Pupil Book 3 page 91**.
* Work through questions 1 and 2 on **Pupil Book 3 page 91** and **page 92** with the children.
* Problem solving: The aim of question 3 is for the children to start thinking about half of an amount, which will be formally presented in the next lesson.

Follow-up

The children can use **Workbook 3 page 74** for independent practice in working with fractions that make up one whole.

Challenge

The children can extend their work by finding fractions of measures such as lengths and masses.

Support

* For some children, question 3 will be too difficult at this point. This is fine, as long as they can identify a fraction of a whole shape or object, and a fraction of a group.
* For the children who need extra practice with the idea of equal sharing, focus on using real objects that can be cut or broken into equal pieces.
* Give the children plenty of practice with fractions with low denominators. For example, you can divide paper plates or circles into quarters, thirds and fifths, and ask the children to label the parts with the correct fractions. You could prepare some sticky note labels for this.
* The children can also start working with tenths, as they have already worked with scales divided into tenths, and this is important preparation for their work on tenths in this unit.
* Make sure the children are confident with these fractions before moving on to ninths, twelfths and fractions with greater denominators.

Interesting mistakes

When the children are exploring the relationship between division and fractions, they may make statements such as: '1 divided by 2 equals 2 because there are 2 halves in 1'. To explore this mistake, you may need to refer back to whole-number division. For example, you could look at $6 \div 2$ and approach it in two different ways:

* *Let's divide 6 into 2 equal groups. How many are in each group?* Set out six counters and split them into two equal groups. *There are 3 in each group.*
  Show that we can do this with the strip of card:
  Let's divide the whole strip into 2 equal parts. How much is in each part? … Each part is half of 1.
* The other way of phrasing the question is: *How many twos can we take away from 6?* Set out six counters and count off pairs of counters. *There are 3 pairs.*
  Then relate this to the whole strip divided into 2 parts. Ask: *How many twos can we take away from 1? We cannot take a whole 2 from 1. 1 is half of 2. We can make half a 2 from 1. So 1 divided by 2 equals one-half.*

Answers for Pupil Book 3 pages 91–92

1   a $\frac{3}{6}$ or $\frac{1}{2}$ b $\frac{1}{4}$ c $\frac{6}{10}$

    d $\frac{2}{3}$ e $\frac{1}{2}$ f $\frac{2}{8}$ or $\frac{1}{4}$

2   a $\frac{4}{8}$ or $\frac{1}{2}$ b $\frac{3}{4}$ c $\frac{1}{6}$ d $\frac{3}{10}$

    e $\frac{6}{12}$ or $\frac{1}{2}$ f $\frac{1}{6}$ g $\frac{1}{9}$ h $\frac{1}{4}$

3 $\frac{4}{24}$ or $\frac{1}{6}$

Half of an amount

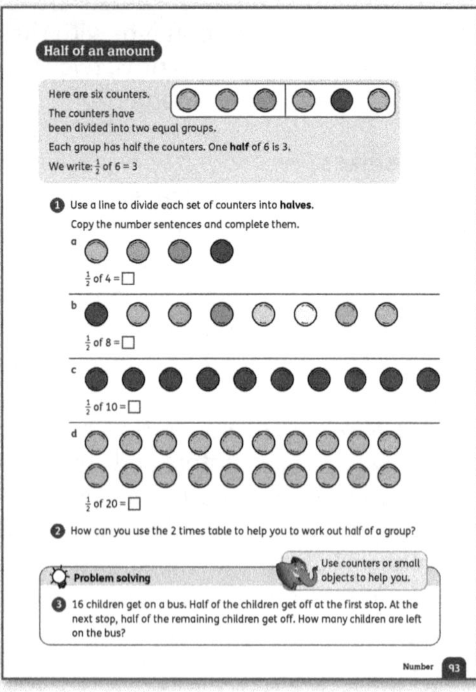

Materials

Small objects such as counters, beans or buttons

Warm-up

- Lay out a set of 12 objects. Ask: *What is half of 12?* (6) Have the children split the set into halves.
- Ask: *How many are there in each half?* (6) *How much do two halves make?* (2 sixes make 12).
- Emphasise that the whole set is 12, and that finding half of the set is the same as dividing by 2.
- Lay out other even numbers of objects and have the children divide them in half, asking questions in a variety of formats, but always writing the mathematical sentence to show the notation:
  - *Half of 10 is equal to … ?* (5) $\frac{1}{2}$ of 10 = ___ (5)
  - *How much is half of 20?* (10) $\frac{1}{2}$ of 20 = ___ (10)
  - *Half of 16 equals how much?* (8) $\frac{1}{2}$ of 16 = ___ (8)

- Then change the order of the question to ask:
  - $3 = \frac{1}{2}$ of ___ (6)
  - $4 = \frac{1}{2}$ of ___ (8), and so on.
- Lay out the set of 12 objects again and ask the class: *How can we find one-third of 12?* Guide them to see that we need to make 3 equal groups.
- Demonstrate using the counters, then invite the children to find two-thirds and three-thirds. They should be starting to notice that when the numerator and denominator are the same, the fraction is equal to 1.

Focus

- Once the children have had plenty of practice working with objects, let them work independently through questions 1–3 on **Pupil Book 3 page 93**.
- Problem solving: Question 3 presents a word problem for the children to solve. Some children may benefit from using counters or small objects to solve this problem.

Challenge

Set the children this challenge: *When the numerator and denominator are the same, the fraction is equal to 1. How can you show this using division?*

Support

Make sure the children can divide sets of objects in half before they move on to other fractions.

Half of an odd number

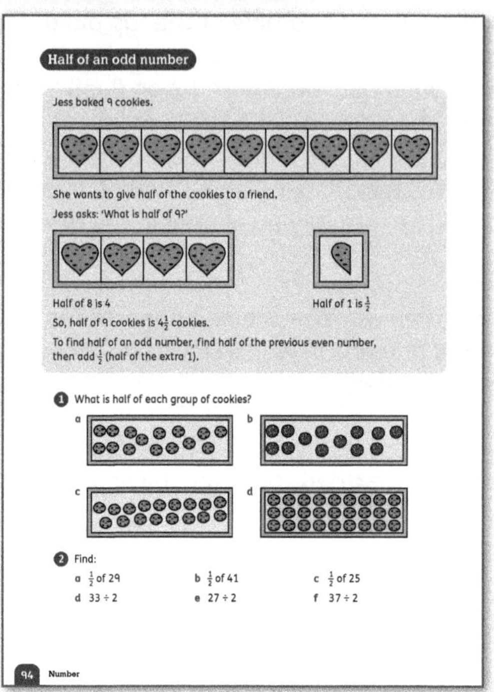

Materials

Cut-out circles to represent cookies

Warm-up

Give each group of children an odd number of cut-out circles. Tell the children that these represent cookies. Ask how they could share the cookies equally between two people. Encourage them to see that they can cut the 'odd' cookie in half. Discuss whether we could use the same method for sharing 7 dolls between 2 children. (No, because we cannot cut a doll in half.)

Focus

Talk through the example on **Pupil Book 3 page 94** and make sure the children know how to write a half as a fraction. Then have them work through questions 1 and 2.

Answers for Pupil Book 3 page 94

1 a $7\frac{1}{2}$ b $6\frac{1}{2}$ c $8\frac{1}{2}$ d $13\frac{1}{2}$

2 a $14\frac{1}{2}$ b $20\frac{1}{2}$ c $12\frac{1}{2}$

  d $16\frac{1}{2}$ e $13\frac{1}{2}$ f $18\frac{1}{2}$

Find fractions

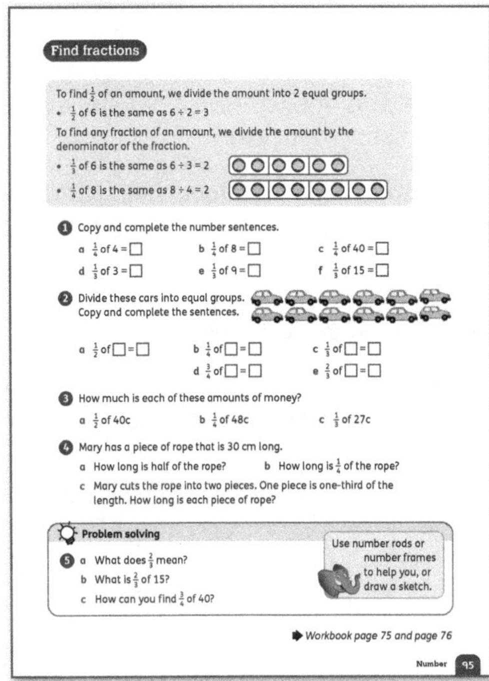

Materials

Counters; large times tables and pictures of arrays for display

Warm-up

- Give each pair of children 12 counters and ask them to divide them into 2 equal rows. Ask: *What fraction of the counters are in each row?* $\left(\frac{1}{2}\right)$
- Repeat for 3 equal rows $\left(\frac{1}{3}\right.$ in each row$\left.\right)$ and 4 equal rows $\left(\frac{1}{4}\right.$ in each row$\left.\right)$.

Focus

- **Pupil Book 3 page 95** extends the work on finding fractions of amounts. Work through the examples with the class.
- Remind the children that they know the times tables facts for small numbers and they can use these facts to find the fractions.
- Let the children complete questions 1–5 independently or in pairs.
- <u>Problem solving:</u> Question 5 gives the children a chance to work with non-unit fractions.

Follow-up

Workbook 3 page 75 and **page 76** provide additional practice in finding fractions of amounts.

Support

Display large multiplication arrays and times tables on the classroom wall. Make sure the children refer to these to help with their calculations.

Interesting mistakes

If any children make statements such as 'Two-thirds equal 6 because $2 \times 3 = 6$', invite them to discuss what the mistake is, and why it is an easy one to make. Here are some questions for you to guide the discussion:

- *Imagine I have a bar of chocolate. I have to divide it between three children.*
  Will each child get more or less than a whole bar? (less)
- *How much will each child get?* (one-third)
- *I have given a third to one child, a third to the second child, but I haven't yet given away the last third. How much have I given away?* (two-thirds)

As an interesting extension, give the children this question:

- *We can say that $\frac{2}{3}$ of $\square = 6$.*
- *Imagine that I have a set of objects and I divide them into 3 equal parts. When I count out 2 of the 3 equal parts, they make 6.*
- *How much is 1 of the parts equal to?* (3)
- *How much are 3 of the parts equal to?* (9)

Answers for Pupil Book 3 book 95

1 a $\frac{1}{4}$ of 4 = 1   b $\frac{1}{4}$ of 8 = 2   c $\frac{1}{4}$ of 40 = 10

  d $\frac{1}{3}$ of 3 = 1   e $\frac{1}{3}$ of 9 = 3   f $\frac{1}{3}$ of 15 = 5

2 a $\frac{1}{2}$ of 12 = 6   b $\frac{1}{4}$ of 12 = 3   c $\frac{1}{3}$ of 12 = 4

  d $\frac{3}{4}$ of 12 = 9   e $\frac{2}{3}$ of 12 = 8

3 a 20c   b 12c   c 9c

4 a 15 cm   b $7\frac{1}{2}$ cm

  c 10 cm and 20 cm

5 a 2 out of 3 equal parts

  b 10   c 40 ÷ 4 = 10, then 3 × 10 = 30

Answers for Workbook 3 page 75

1 a $\frac{1}{2}$ of the squares = 12   b $\frac{1}{2}$ of the squares = 6

   $\frac{1}{3}$ of the squares = 8 $\frac{1}{3}$ of the squares = 4

   $\frac{1}{4}$ of the squares = 4 $\frac{1}{4}$ of the squares = 3

   c $\frac{1}{2}$ of the squares = 8

   $\frac{1}{4}$ of the squares = 6

2 a 9   b 10   c 16
   d 24   2 3   f 36

Answers for Workbook 3 page 76

1 a 5 bottles shaded   b 3 ladybirds with spots
   c 2 purple cars   d 2 green fish
   e 2 yellow birds   f 4 striped beads

2 Individual answers

Equivalent fractions

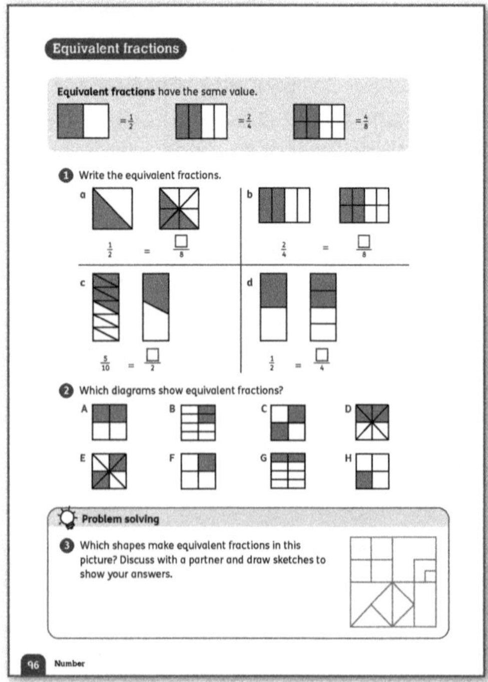

Materials

Paper plates or circles cut from paper or card, divided into a variety of equivalent fractions: thirds and sixths, halves, quarters and eighths, fifths and tenths; fraction wall (you can search for this online)

Warm-up

Show the children the paper plates or circles divided into fractions. Show them a half and ask them to find other fractions that look the same size. You can ask questions such as:

- *Which other fractions are the same size and shape as a half?*
- *How many quarters make a half?*
- *How many sixths make a half?*
- *What other fractions make a half?*

Focus

- Introduce the term *equivalent fraction*. Ask the children if they notice any part of the word *equivalent* that looks familiar. They may notice that 'equi-' looks similar to 'equal'. Explain that this is because equivalent fractions are equal in size.
- You can also demonstrate this using fractions of sets of objects. Set out 12 objects and ask:
  - What is one-third of 12? (4)
  - What is one-sixth of 12? (2)
  - What is two-sixths of 12? (4)
  - What can we say about two-sixths and one-third? (They are equivalent.)
- You can repeat this approach with different equivalent fractions ($\frac{2}{4}$ and $\frac{1}{2}$; $\frac{3}{4}$ and $\frac{6}{8}$; and so on) and different numbers of objects.
- Spend some time working through some practical examples, then go through the example at the top of **Pupil Book 3 page 96**.
- The children can complete questions 1 and 2 independently.
- <u>Problem solving</u>: You might want the children to do question 3 in pairs or groups. The children can approach this question according to their own level of understanding: some children may focus on halves and quarters; others may find more complex fractions.

Support

Some children may find the problem-solving question too challenging. It is fine to leave it out if it is not at an appropriate level for your class.

Some children may need additional help with identifying fractions that are the same size in shaded shapes. Make sure the shapes are all the same size and shape.

Answers for Pupil Book 3 page 96

1 a $\frac{1}{2} = \frac{4}{8}$ b $\frac{2}{4} = \frac{4}{8}$ c $\frac{5}{10} = \frac{1}{2}$ d $\frac{1}{2} = \frac{2}{4}$

2 A, C, D and E all have $\frac{1}{2}$ shaded. B, F, G and H all have $\frac{1}{4}$ shaded.

3 There are many possible answers, including:

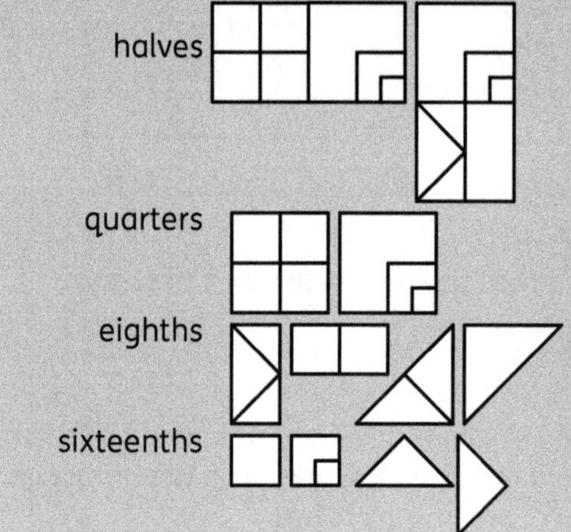

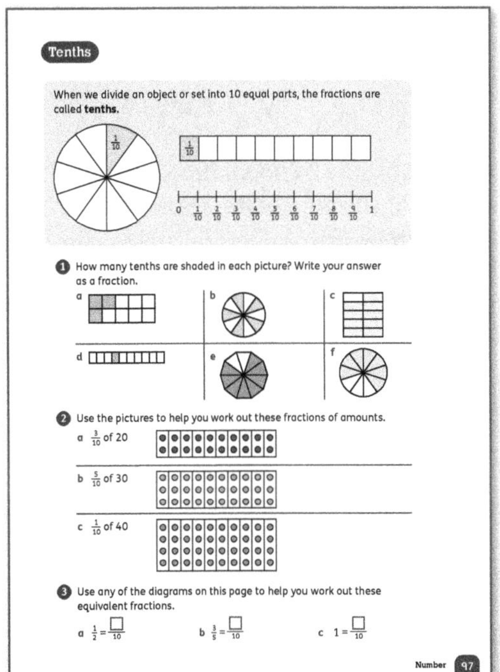

Materials
Large number line from 0 to 1, divided into tenths

Warm-up
- Show a number line from 0 to 1, with unlabelled marks for the tenths:

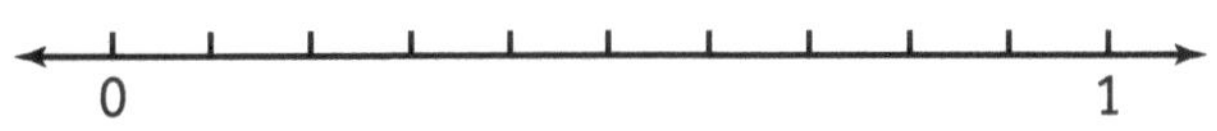

- Ask:
  - What does this number line show?
  - Which whole numbers can you see? (0 and 1)
  - What do the marks along the line show us?
  - How many parts are there between 0 and 1? (10)
  - What do we call each part? (Tenths)

Focus
- Remind the class that we can divide whole numbers into smaller parts. Label the marks on the number line in tenths:

$$\frac{1}{10} \quad \frac{2}{10} \quad \frac{3}{10} \quad \frac{4}{10} \quad \frac{5}{10} \quad \frac{6}{10} \quad \frac{7}{10} \quad \frac{8}{10} \quad \frac{9}{10}$$

- Let the children count in tenths with you from 0 to 1. Explain that $\frac{10}{10}$ is the same as 1. *When we have counted all 10 parts, we have one whole.*
- Ask questions about the number line to reinforce understanding:
  - Show me $\frac{1}{10}$ on the number line.
  - Which fraction is halfway between 0 and 1? (five-tenths) Do you know another name for this fraction? (one-half)
  - Point to $\frac{3}{10}$ on the number line. Count on three more tenths. Which fraction are you on? How many more tenths will make one whole?
- Continue with similar questions.

- Discuss questions 1–3 on **Pupil Book 3 page 97** before letting the children work through them independently.

Answers for Pupil Book 3 page 97
1 a $\frac{3}{10}$ b $\frac{5}{10}$ c $\frac{9}{10}$

    d $\frac{1}{10}$ e $\frac{8}{10}$ f $\frac{7}{10}$

2 a 6   b 15   c 4

3 a $\frac{1}{2} = \frac{5}{10}$ b $\frac{3}{5} = \frac{6}{10}$ c $1 = \frac{10}{10}$

Compare fractions

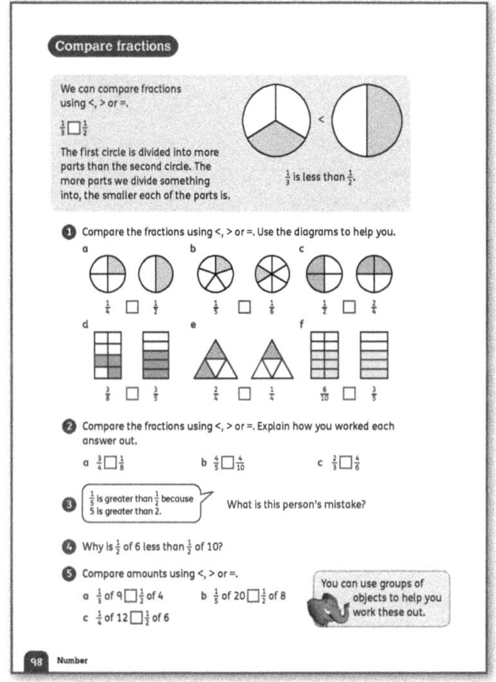

Materials
Counters or other small objects

Warm-up
Revise the < and > signs (*greater than* and *less than* signs) with the class. If necessary, do some number comparisons using whole numbers before you move on to fractions.

Focus
- Work through the example at the top of **Pupil Book 3 page 98**.
- Work with the class through the first few parts of question 1 before letting them work independently to complete the rest of question 1 and questions 2 and 3 independently.
- For questions 4 and 5, the children can use groups of objects such as counters to help them work out the fractions of amounts.

Challenge
Pose more challenging comparisons of fractions of amounts, for example: *Which is greater: $\frac{3}{4}$ of 20 or $\frac{1}{2}$ of 24?* $\left(\frac{3}{4}\ of\ 20\right)$; $\frac{1}{4}$ of 100 or $\frac{1}{2}$ of 60? $\left(\frac{1}{2}\ of\ 60\right)$

Support

Assist the children to make fractions of amounts using small objects or counters.

Interesting mistakes

Some children may make statements such as '$\frac{1}{3}$ of 9 > $\frac{1}{2}$ of 4 because 9 is greater than 4.' The > sign is correct here but the reasoning is not.

Discuss these types of errors with the class and ask the children to explain the mistake and how to correct it. Emphasise that we need to compare the fraction of the quantity, not the quantity by itself.

Answers for Pupil Book 3 page 98

1. a $\frac{1}{4} < \frac{1}{2}$ b $\frac{1}{5} > \frac{1}{6}$ c $\frac{1}{2} = \frac{2}{4}$
   d $\frac{3}{8} < \frac{3}{5}$ e $\frac{2}{4} > \frac{1}{4}$ f $\frac{6}{10} = \frac{3}{5}$

2. a $\frac{3}{4} > \frac{1}{8}$ b $\frac{4}{5} > \frac{4}{10}$ c $\frac{2}{3} = \frac{4}{6}$
   Possible explanation: Write the fractions with the same denominator to make them easier to compare. For example, in part a, change $\frac{3}{4}$ to $\frac{6}{8}$. Then compare the numerators.

3. This person has confused fractions with whole numbers. The denominator (number at the bottom of the fraction) tells us how many parts the fraction is divided into. The greater that number, the smaller each of the parts.

4. Possible answer: 6 is less than 10, so $\frac{1}{2}$ of 6 is less than $\frac{1}{2}$ of 10.
   or $\frac{1}{2}$ of 6 = 3, $\frac{1}{2}$ of 10 = 5. 3 < 5

5. a $\frac{1}{3}$ of 9 (3) > $\frac{1}{2}$ of 4 (2)   b $\frac{1}{5}$ of 20 (4) = $\frac{1}{2}$ of 8 (4)
   c $\frac{1}{4}$ of 12 (3) = $\frac{1}{2}$ of 6 (3)

Fractions of a number

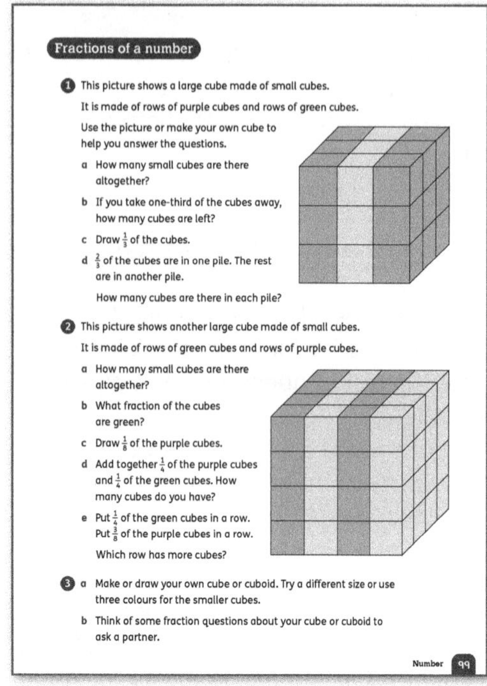

Materials

Small objects to sort into sets; egg boxes; cubes

Warm-up

- Draw a simple picture of a box of 12 chocolates, arranged in a 3 × 4 array. Give the children this problem:
  - *I have a box of 12 chocolates. I have eaten three-quarters of the chocolates. How many chocolates have I eaten?* (9)
- Let the children make guesses or suggestions. Let them come to the board and demonstrate how to work it out.
- Ask questions such as:
  - *Is three-quarters more or less than half?*
  - *How many quarters are there in half the box?*
  - *How much is one-quarter of 12?*
- Let the children use their own reasoning to suggest how to work out three-quarters of the chocolates (for example, work out one-quarter of 12 and use that answer to find two-quarters and then three-quarters).

Focus

- Give the children other, similar problems. For example, draw a row of 15 beads on a string and say: *I want to colour $\frac{2}{3}$ of the beads. How can we work out how many beads to colour?*
- Make sets of a given number of small objects and ask the children to make a given fraction of the objects.
- Give each group of children one or more egg boxes (for example, a box that holds 12 or 18 eggs) and ask them to put objects into the egg box cups to show a given fraction $\left(\frac{1}{3}, \frac{5}{6}, \text{etc.}\right)$.
- Discuss the pictures of cubes on **Pupil Book 3 page 99** and ask the children to work through questions 1–3 on the page.

Support

Any children who need to can use real cubes to help them answer the questions.

Answers for Pupil Book 3 page 99

1. a 27   b 18
   c drawing of 9 cubes   d 18 and 9
2. a 64   b $\frac{1}{2}$
   c drawing of 4 purple cubes   d 16
   e the purple row
3. Individual answers

Add and subtract fractions

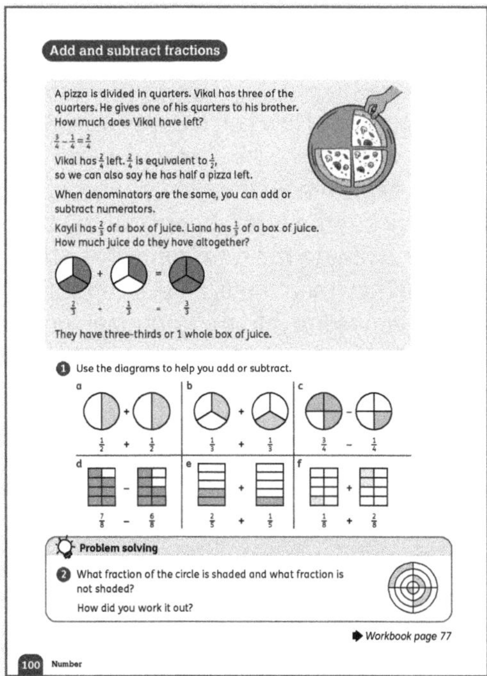

Materials
Squared paper

Warm-up
Give each child a small piece of squared paper, for example a rectangle of 4 × 10 blocks. Instruct them to colour fractional parts. For example: *Colour half the rectangle red. Colour one-tenth of the rectangle green.*

Focus
- Teach addition and subtraction of fractions using similar examples to those given at the top of **Pupil Book 3 page 100**. Use diagrams of shapes divided into fractional parts to help with understanding.
- At this level, the children only add and subtract fractions with the same denominator, so the calculations only involve the numerators.
- When the children have had sufficient practice, they can complete question 1.
- Problem solving: In question 2 the children have to visualise moving all the shaded areas round to one of the quarters. Alternatively, they may notice that one-quarter of each ring is shaded, and since the four parts of each ring are identical, $\frac{1}{4}$ of four rings is still $\frac{1}{4}$ of the whole shape. They then have to subtract the shaded quarter from the whole to work out the fraction not shaded.

Follow-up
Workbook 3 page 77 provides extra practice and consolidation. The children start by shading diagrams. Then they move on to addition and subtraction number sentences and finally they solve word problems.

Answers for Pupil Book 3 page 100

1 a 1   b $\frac{2}{3}$ c $\frac{2}{4}$ d $\frac{1}{8}$ e $\frac{3}{5}$ f $\frac{3}{8}$

2 $\frac{1}{4}$ shaded, $\frac{3}{4}$ not shaded
Possible answer: I moved the shaded sections and put them all in the same quarter of the circle.

Answers for Workbook 3 page 77

1 a $\frac{3}{4}$ b $\frac{5}{6}$ c $\frac{3}{5}$

2 a $\frac{2}{3}$ b $\frac{3}{5}$ c $\frac{6}{10}$ d $\frac{1}{8}$ e $\frac{7}{10}$ f $\frac{1}{2}$

3 a Correct. $\frac{8}{9} + \frac{1}{9} = \frac{9}{9}$, which is the same as 1.
   b Incorrect. $\frac{4}{5} - \frac{1}{5} = \frac{3}{5}$
   c Correct. Any number subtracted from itself is 0.
   d Incorrect. $\frac{3}{4} - \frac{1}{2} = \frac{3}{4} - \frac{2}{4} = \frac{1}{4}$

4 $\frac{7}{10}$ m

End-of-unit check

Assess the work done on fractions by asking questions such as:
- *How many halves make a whole?* (2)
- *How many quarters make a whole?* (4)
- *A pizza is divided into 5 equal slices. What fraction of the pizza is each slice?* (one-fifth)
- *We want to break a bar of chocolate into equal pieces so that some children can have one-tenth each. How many equal parts must we break the chocolate into?* (10)
- *Can you name another fraction that is equal to four-sixths?* (two-thirds)
- *How many tenths make a half?* (5)
- *A pizza is divided into 6 equal parts. What fraction of the pizza is 2 parts?* (one-third)
- *Is three-fifths greater than, less than or equivalent to one-half?* (greater than)
- *What is one-quarter of 16?* (4)
- *How can we share five apples fairly between two people?* $\left(2\frac{1}{2} \text{ apples each}\right)$
- *What is half of 9?* $\left(4\frac{1}{2}\right)$
- Show a number line divided into tenths, with $\frac{1}{10}$ labelled. Point to one of the divisions on the line and ask: *How many tenths is this? What is $\frac{1}{10} + \frac{2}{10}$?* $\left(\frac{3}{10}\right)$

Learning objectives
- Estimate, measure, compare, add and subtract volume/capacity in litres and millilitres.
- Use instruments that measure capacity (ℓ/ml) and temperature:
  - estimate first and then measure
  - round up if the measurement is past or equal to the halfway point between readings
  - round down if the measurement is less than the halfway point between readings.
  - use instruments that measure temperature.

Key words
capacity   full   half full   litre   millilitre   temperature   thermometer   degree   hot   warm   cool   degrees Celsius (°C)

Unit introduction

Materials
A variety of containers marked with their capacities in litres or millilitres (drinks bottles, cartons, jugs, etc.); a variety of containers that are not marked with their capacities (cups, beakers, bowls, empty bottles, food pots and tubs, etc.); 1-litre bottles (unmarked), marker pens, water or sand

Teaching guidance
Before starting this unit, ask the children to:
- Look at information labels on various containers at home and look for the word 'litre'.
- Collect and bring to school a variety of bottles and containers, such as empty milk bottles, juice bottles, shampoo and shower gel bottles, yoghurt pots and margarine tubs. They will use these containers in the estimation and measuring activities in this unit. The children can also write down which containers they found, and how many litres or millilitres each one holds.

As a mental maths warm-up, to prepare the children for working with millilitres and litres, ask the children to find pairs of multiples of 100 that make a total of 1000.

As a practical introductory activity for this unit, the children could estimate and find the capacities of various containers. You will need to prepare some 1-litre containers, such as juice cartons, by marking and labelling them at the $\frac{1}{4}$-litre and $\frac{1}{2}$-litre levels.

The children can then carry out a survey of the capacities of containers used for different food and drink products:
- First give each group one of your prepared 1-litre containers. Say: *I want you to look at your containers. Can you see the marks? This mark is where the*

container is one-quarter full. This mark is where the container is half full. *How can we use these marks to work out the capacity of the container when it is* full? Let the children explain how the marks help us, and how they can work out the total capacity of the container. They should estimate the total capacity to the nearest quarter litre. They then measure the capacity by pouring water or sand into the container, and pouring it into a marked measuring jug. Ask: *How accurate was your estimate? Why do you think it was/wasn't close to the actual capacity? What helped you to estimate?*
- Give the groups some smaller containers and ask them to use the capacity of the 1-litre container to estimate the capacity of each container to the nearest $\frac{1}{4}$ litre. They can then fill each container and pour into a measuring jug to check their estimates. Some of the containers may hold exactly $\frac{1}{4}$ litre or $\frac{1}{2}$ litre, but it is likely that the majority will not. Introduce the idea that we need a smaller unit of capacity and introduce the millilitre.

Capacity is similar to mass in that 1000 ml = 1 *litre*. You can point out this similarity to the children (*milli*- is from the Latin word for thousand). Tell the children that there are 1000 ml in 1 litre and the symbol for the *millilitre* is 'ml'. Use a calibrated container, such as a measuring jug, to demonstrate that 1 litre = 1000 ml, $\frac{1}{2}$ litre = 500 ml and $\frac{1}{4}$ litre = 250 ml.

Measure capacity

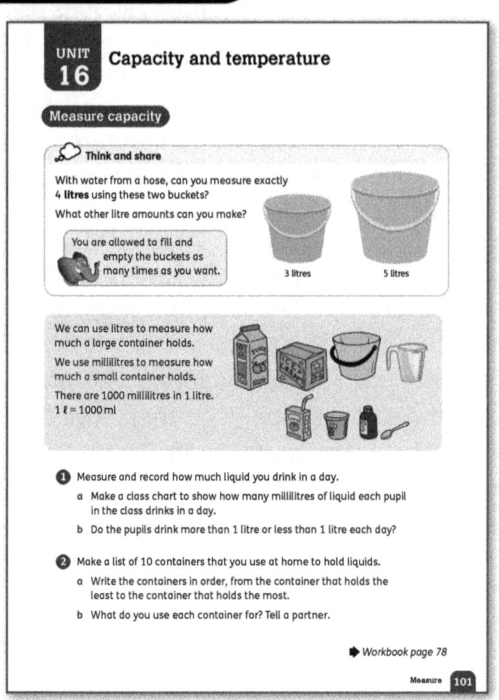

Materials
A variety of containers with a capacity of less than 1 litre; 1-litre bottle, half-litre bottle, measuring spoons, small measuring cups, 1-litre measuring jug

Warm-up

- <u>Think and share:</u> Work through this problem-solving activity on **Pupil Book 3 page 101**.
- Discuss the problem using the number talk structure (pages 17–18):
  - Present the problem, and let the children volunteer possible ideas for how to work out answers.
  - Do not suggest a method for working it out – let the children develop their own strategies.

Focus

Questions 1 and 2 on **Pupil Book 3 page 101** involve practical activities, some of which will need to be set as homework. Decide how you will manage this beforehand. For example, give the children a simple chart and ask them to tick one block each time they drink a glass of water or other drink during one whole day.

Follow-up

Once the children have done the practical measuring activities, let them work through **Workbook 3 page 78** to show their understanding of litres and millilitres. Alternatively, the children can complete this activity after their work on reading scales on **Pupil Book 3 page 103**.

Challenge

The 'Think and share' problem is a type of water pouring puzzle. You can find similar puzzles online. Here is a classic problem:

I have an 8-litre container (A) that is full of water. I also have two smaller containers that are empty: a 5-litre container (B) and a 3-litre container (C). How can you pour the water so that you have exactly 4 litres in one container and 4 litres in another?

Rules:
- *None of the containers have markings on them, so you cannot work out any fractional amounts.*
- *You have no other measuring containers.*

Let the children work on the problem in pairs. They should draw sketches or diagrams to explore how to pour the water from one container to another in order to measure the required amounts. Do not give the solution away.

These are the steps to solve this puzzle:
- Step 1: Pour water from container A to fill container B, leaving 3 litres of water in container A and 5 litres in container B.
- Step 2: Pour water from container B to fill container C. We now have 3 litres of water in A, 2 litres of water in B and 3 litres of water in C.
- Step 3: Empty C into A. A now contains 6 litres of water.
- Step 4: Pour 2 litres of water from B into the empty C.
- Step 5: Pour water from A (which now contains 6 litres) to fill the empty B. We now have 1 litre of water in A, 5 litres of water in B and 2 litres of water in C.
- Step 6: Pour water from B to fill C (which already contains 2 litres of water, so you are adding 1 litre from B).
- Step 7: We are left with 4 litres of water in B. We have 3 litres of water in C: pour this into A, which already contains 1 litre of water. This gives us 4 litres of water in A and B.

Support

Some children may need extra practice in comparing containers and identifying how much they hold. Allow the children who are finding the work challenging to work with two or three containers. Ask questions such as:
- *What (food or drink) do you usually get in this container?*
- *Can you pour it?*
- *When it is full, before any is poured out or used, how much does it hold?*
- *Does it look like it holds more or less than this bottle?* (Show a 1-litre bottle.)
- *Does it look like it holds more or less than this bottle?* (Show a half-litre bottle.)

Demonstrate the concept of millilitres, using measuring spoons and small measuring cups to show 5 ml, 10 ml, 50 ml, 100 ml, building up to 1000 ml.

Fill a 1-litre jug slowly in 100-ml increments, counting in hundreds.

Answers for Pupil Book 3 page 101

<u>Think and share:</u> Fill the 3-litre bucket and empty it into to 5-litre bucket. Fill the 3-litre bucket again and add water to the 5-litre bucket to fill it. You will have 1 litre left in the 3-litre bucket.

Then empty the 5-litre bucket and fill it with the 1 litre from the 3-litre bucket. Fill the 3-litre bucket again and empty it into the 5-litre bucket. Now there are 4 litres in the 5-litre bucket.

1 litre, 2 litres, 3 litres, 5 litres, and 8 litres.

1–**2** Individual answers

Answers for Workbook 3 page 78

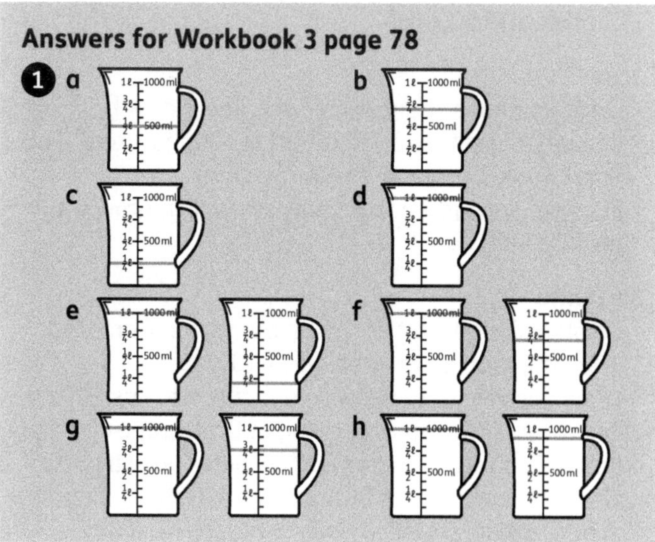

Litres and millilitres

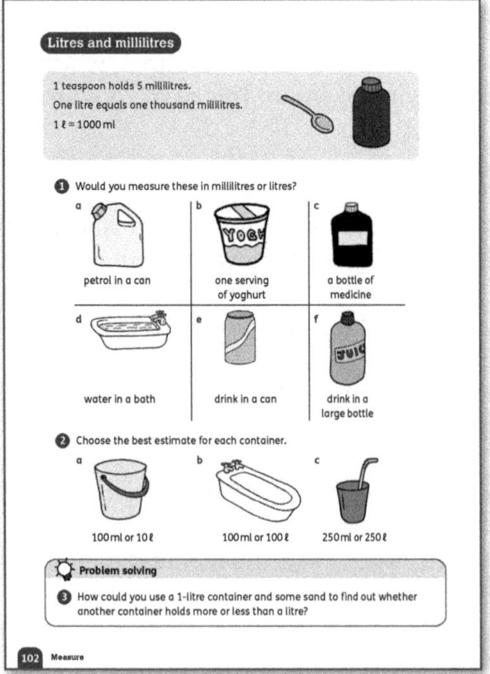

Materials
Measuring spoons and cups; measuring jug, stones

Warm-up
- Show the different measuring spoons and cups to the class, and let the children say which holds the most and which holds the least.
- Hold up one of the spoons or cups and have the children identify how many millilitres it holds.
- Ask questions that lead into an understanding of equivalence, for example:
  - How many of these spoons/cups do I need to make 100 ml?
  - (If the spoon/cup holds an amount that is not a factor of 100, for example 30 ml): What other spoon or cup do I need to make exactly 100 ml?
  - I want to measure 500 ml. Which are the best containers to use?
  - Why is it not a good idea to use this (holding up a teaspoon) to measure 500 ml?
  - How can I use this (holding up a 250-ml cup) to measure 500 ml?

Focus
- Once you have spent some time dealing with equivalent units, work through questions 1 and 2 on **Pupil Book 3 page 102** with the class.
- <u>Problem solving:</u> Let the children work in groups to discuss question 3.

Challenge
- Display a large, vertical measuring scale from 0 to 1000 ml, marked in intervals of 100.
- Draw a line on it to show a capacity (such as 200 ml).
- Tell the children you are going to drop stones into the liquid. Each stone increases the water level by 50 ml.
- Show different groups of stones with a different number in each group. Let the children write the level that would be shown on the scale for each group of stones.

- You could demonstrate how dropping stones into a measuring jug of water increases the level of the water beforehand, if you wish.
- This activity also works well with mass.

Interesting mistakes
- The capacity of a container is the amount it holds when it is full. The children may sometimes think that capacity is the amount a container is holding at a given point.
- You may need to demonstrate capacity practically.
- For example, show a container that has a capacity of 1 litre. Half-fill the container with water and say that the container is now holding half a litre of water, but the capacity of the container is still 1 litre.

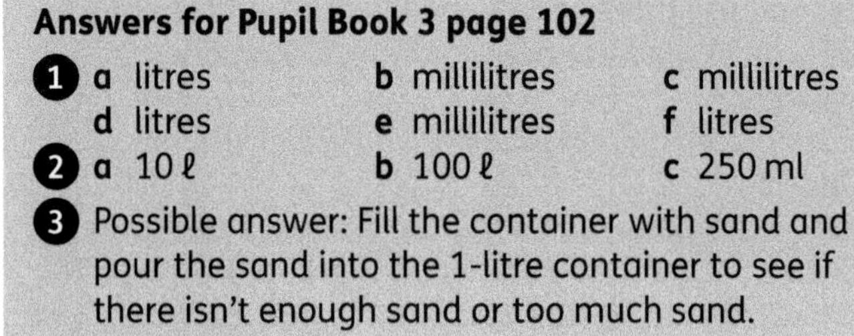

Read scales

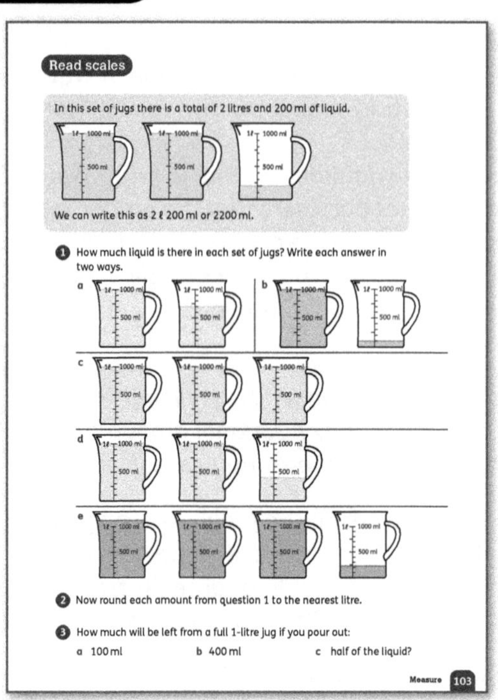

Materials
Measuring jugs

Warm-up
- The children have already read scales when they worked with mass and length.
- Show the children a measuring jug and explain that measuring capacity is rather different from measuring mass, as there is no pointer or other moving parts on the scale of a measuring jug.
- Show the children how to look at where the level of the liquid is on the scale. It is similar to reading a length on a ruler or tape measure.

Focus

- Work through the example at the top of **Pupil Book 3 page 103** and demonstrate the different ways of writing measurements: in litres and millilitres (2 ℓ 200 ml) and in millilitres (2200 ml).
- Begin working through questions 1–3 with the class, then let them continue independently if possible.

Follow-up

If the children have not yet completed **Workbook 3 page 78**, they can do this activity now to consolidate their work on reading scales.

Support

Some children may need practice counting in hundreds to 1000 and reading in hundreds off a calibrated scale.

Answers for Pupil Book 3 page 103

1 **a** 1 ℓ 700 ml, 1700 ml **b** 1 ℓ 100 ml, 1100 ml
   c 3 ℓ, 3000 ml **d** 2 ℓ 400 ml, 2400 ml
   e 3 ℓ 200 ml, 3200 ml
2 **a** 2 ℓ **b** 1 ℓ **c** 3 ℓ **d** 2 ℓ **e** 3 ℓ
3 **a** 900 ml **b** 600 ml **c** 500 ml

Capacity problems

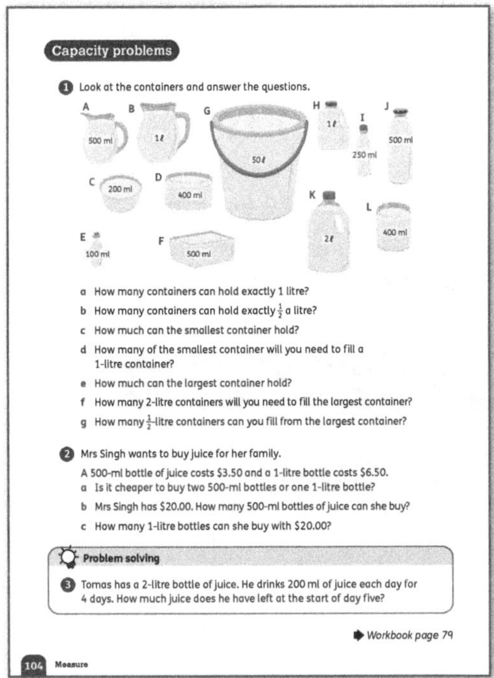

Materials

Cards showing capacities in millilitres and litres

Warm-up

- Give some of the children cards showing capacities in millilitres and litres. Ask the children to arrange themselves in order from least to greatest capacity.
- Choose two of the cards and ask a child to suggest another capacity that would fit in between them.

Focus

- Let the children work in pairs to complete the capacity problems in questions 1–3 on **Pupil Book 3 page 104**. Assist them as needed.
- Problem solving: The children can draw diagrams to help them solve the word problem in question 3.

Follow-up

Use the questions on **Workbook 3 page 79** to consolidate the children's understanding of litres and millilitres.

Support

Some children may need assistance with the calculations required to solve the problems. Let the children help each other to understand each problem, how to work out the answer and then to check that the answer makes sense. They can use a variety of strategies: repeated addition, partitioning, and so on.

Interesting mistakes

- The children might get confused when comparing different units. For example, they might say: '50 ℓ is less than 500 ml because 50 is less than 500.'
- Remind them to work out how many millilitres there are in 1 litre, then how many litres there are in 50 litres. Encourage them to do this mentally.

Answers for Pupil Book 3 page 104

1 **a** B, H **b** A, J, F **c** 100 ml **d** 10
   e 50 ℓ **f** 25 **g** 100
2 **a** One 1-litre bottle **b** 5 **c** 3
3 1 ℓ 200 ml or 1200 ml

Answers for Workbook 3 page 79

1 From the bottom of the jug: 100 ml, 200 ml, 300 ml, 500 ml, 600 ml, 700 ml, 800 ml
2 **a** 1 ℓ **b** 500 ml **c** 250 ml
3 Possible answers: Fill container A 20 times. Fill container B 10 times. Fill container C and container D once each. Fill container B 5 times and container C 2 times.
4 **a** 80 ℓ **b** 150 ml **c** 370 ml

Temperature

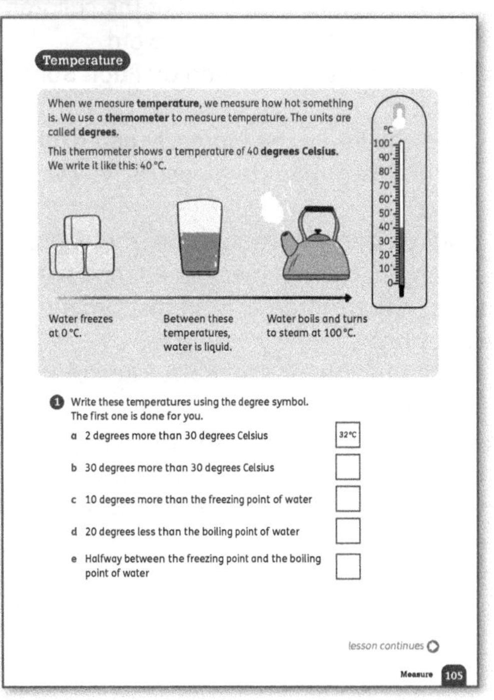

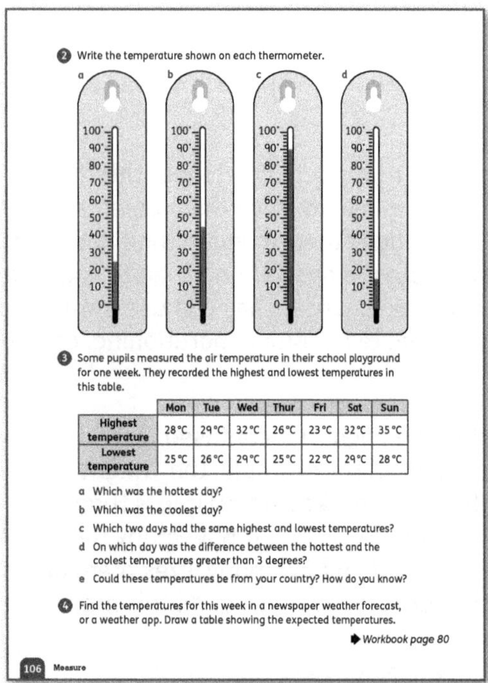

Materials

Thermometer; large model or drawing of a thermometer

Warm-up

- Draw a picture of an ice cream on the board. Ask:
  - What can you see?
  - Is ice cream warm or is it cold?
  - Who can tell me something that is colder than ice cream?
  - Can you tell me something warmer than ice cream?
  - What is warmer than that?
- Let the children keep suggesting a variety of cold, *cool*, *warm* and *hot* items. Write them on the board and let the children compare the items: which are warmer and which are cooler.

Focus

- Write the word *temperature* on the board. Ask if any children know what it means. Let the children share their own understanding of the word.
- Work through the information on **Pupil Book 3 page 105**.
- Show the children a *thermometer* and explain that we use this instrument to measure how warm or cold things are.
- The children may notice that the *-meter* part is a unit of length. *Thermo* means heat and *meter* means measure. A thermometer is an instrument that measures heat.
- Introduce the unit we use to measure temperature: *degrees Celsius (°C)*.
- Demonstrate how to write the degree symbol correctly and talk about the temperature measurements on **Pupil Book 3 page 105** and **page 106**.
- Work through questions 1–4 on the two pages together. Some children may prefer to complete the activities independently.

Follow-up

You can use **Workbook 3 page 80** to consolidate this work.

Challenge

Some children may like to find out the temperatures on other planets. These temperatures range from negative numbers to thousands of degrees Celsius. The children can research this independently.

Support

Give the children practice in reading the scale of a model thermometer or a drawing of a thermometer. Give them practice in comparing temperatures and saying which temperature is warmer/hotter and which is cooler/colder. This will reinforce the concept that the higher the number the hotter the temperature.

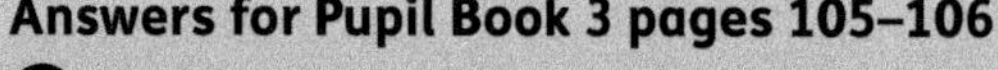

Answers for Pupil Book 3 pages 105–106

1. a 32 °C (Provided as an example)
   b 60 °C   c 10 °C
   d 80 °C   e 50 °C
2. a 26 °C (Provided as an example)   b 46 °C
   c 90 °C   d 16 °C
3. a Sunday   b Friday
   c Wednesday and Saturday
   d Sunday
   e Individual answers depending on the country
4. Individual answers

Answers for Workbook 3 page 80

1. a–f Thermometer scales coloured to match the temperatures given.

Temperature experiments

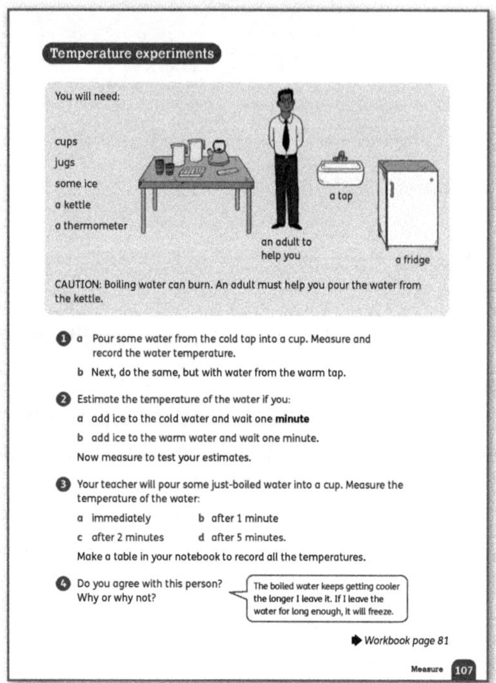

Materials

Cups, jugs, kettle, electricity source, ice, thermometer, smartphone or stopwatch for timing the intervals

Warm-up

* Before the lesson, prepare all the materials you will need for the practical activity (see the Materials section above).
* At the beginning of the lesson, have the children copy this table into their notebooks:

			Estimate	Measurement
1	a	water from cold tap		
	b	water from warm tap		
2	a	cold water and ice after 1 minute		
	b	warm water and ice after 1 minute		
3	a	just-boiled water		
	b	boiled water after 1 minute		
	c	boiled water after 2 minutes		
	d	boiled water after 5 minutes		

Focus

Read through steps 1–4 on **Pupil Book 3 page 107** with the class so that they understand what is going to happen and what you are measuring.

Safety note: It is best to carry out this activity as a demonstration to the class. Allow some children to be your assistants: pouring water from the cold and warm taps, adding ice, measuring the temperature, timing the intervals, and so on. However, you must not allow them to handle the kettle or pour the boiling water. You should measure the temperature of the just-boiled water after each interval, as even after 5 minutes, it may still be hot enough to scald.

Follow-up

Workbook 3 page 81 provides practice with both temperature and reading information from a bar chart.

Answers for Pupil Book 3 page 107

1–**3** Individual answers based on practical work

4 No. The temperature will not go below room temperature. It will not freeze unless it is in a place where the air temperature is below freezing.

Answers for Workbook 3 page 81

1 a Temperatures in New Delhi, India, for 12 months
  b January, February, December
  c May, June, July
2 a March, April, September, October
  b May, June, July
3 Individual answers. For example: January 15 °C, April 28 °C, November 20 °C, December 16 °C
4 Individual answers. For example: shorts, T-shirts and sandals because it is very hot

End-of-unit check

To assess the children's understanding of capacity, ask questions such as:

* *Put these capacities in order from smallest to greatest: $\frac{1}{2}$ litre, 300 ml, 1 litre, 750 ml (300 ml, $\frac{1}{2}$ litre, 750 ml, 1 litre)*
* *How many half-litre bottles of lemonade do you need to fill a 4-litre bottle? (8)*
* *4 small bottles of water hold the same amount as one 3-litre bottle. How much water is in a small bottle? (750 ml)*
* *Two different-sized bottles together hold 2 litres of water. If one of the bottles holds $1\frac{1}{4}$ litres, how much does the other bottle hold? ($\frac{3}{4}$ litre or 750 ml)*
* *What is the equivalent of 500 ml in litres? ($\frac{1}{2}$ litre or 0.5 litre)*
* *Show water in a measuring jug. How many millilitres is this, to the nearest 100 ml?*
* *What units would you use to measure the amount of water in a bucket? (litres) In a cup? (ml)*

To assess the children's understanding of temperature, use a thermometer template. If possible, laminate photocopies of the thermometer template and use a marker pen to draw lines to show a range of temperatures for the children to read.

Ask question such as:

* Give the children two temperatures in degrees Celsius. *Which is warmer/cooler?*
* *Put these temperatures in order from coolest to warmest: 31 °C, 22 °C, 49 °C, 15 °C. (15 °C, 22 °C, 31 °C, 49 °C)*
* *What temperature is 5 degrees cooler than 20 degrees Celsius? (15 °C)*
* *What temperature is 10 degrees warmer than 25 degrees Celsius? (35 °C)*

Learning objectives
- Use informal language and the language of probability to describe the chance of an event happening.
- Conduct probability experiments, and present and describe the results.

Key words
chance possible impossible event definite likely unlikely equal chance outcome probability experiment

Unit introduction

Materials
A simple spinner with four quarters in different colours – red, blue, yellow, grey (You can make a spinner from a circle of card, an arrow cut from card and a paper fastener or split pin.)

Teaching guidance
Show the spinner to the class and ask: *What colour do you think the arrow will land on when you spin it?* Let them guess. Ask questions such as:
- *Who thinks the arrow will land on red?*
- *Who thinks it will land on grey?*
- *Is it possible for the arrow to land on yellow?* (Yes.)
- *My guess is that it will land on a colour that is not red. Is there a good chance of that?* (Yes.) *Why?* (There are three sections that are not red and only one section that is red.)

Name a colour that is not on the spinner, for example black. Ask: *Is it possible for the arrow to land on that colour? Why not?*

Write the words *possible* and *impossible* on the board. Let the children identify possible and impossible colours for the arrow to land on. (Possible: red, blue, yellow, grey. Impossible: any other colour.)

Will it happen?

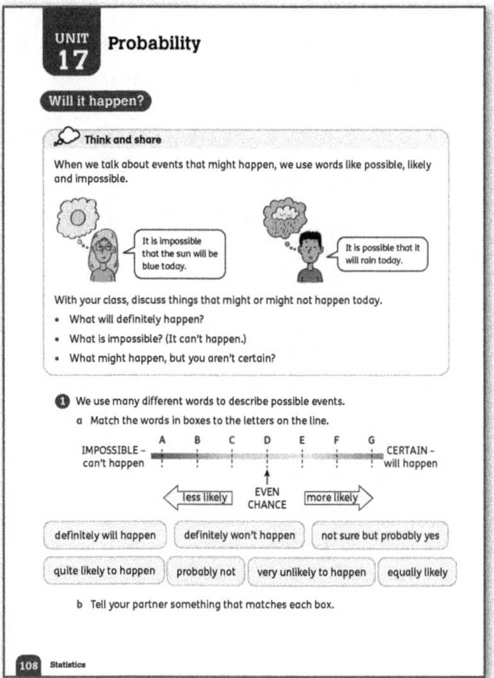

Materials
Simple spinners (made from circles of card, an arrow cut from card, a paper fastener or split pin); strips of paper

Warm-up
- <u>Think and share:</u> Discuss the pictures in this section of **Pupil Book 3 page 108** with the class.
- Draw a horizontal line on the board and label the right-hand end of the line 'Definitely will happen' and the left-hand end 'Definitely will not happen'.
- Let the children say where the possible and impossible *events* shown in the pictures belong on the line.
- Then let them suggest some events that definitely will happen, definitely won't happen, and others that might happen.
- Let them discuss where on the line they think each event should be placed – closer to 'Definitely will happen' or closer to 'Definitely will not happen'.
- Help them to notice that the more certain we are of an event, the further we place it towards the right-hand end of the line; the less certain we are (or the more unlikely we think an event is), the further we place it towards the left-hand end.

Focus
- In question 1, the children discuss the phrases and decide in pairs where to place each phrase on the line from 'Impossible' to 'Certain'. They should copy the line into their notebooks and write each phrase in the place they decide.

- Ask the children to discuss the following events. Ask: *How likely is each event to happen to you?*
  - A unicorn will appear in the school playground.
  - The sun will set this evening.
  - I will go to bed at 4.40 p.m. this afternoon.
  - I will eat rice with my dinner.
  - I will play with my friend after school.
  - I will see the moon tonight.
  - I will see three red cars in a row.

Support

The children may need support with the vocabulary. If they are finding this very challenging, spend some time exploring 'possible' and 'impossible' events on simple spinners.

You can create a poster for the classroom wall with a table of terms and definitions, like this one.

Term	Definition	Examples
definite	certainly will happen	
possible	might happen	
impossible	definitely won't happen	
likely	probably will happen	
unlikely	probably won't happen	
equal chance	just as likely to happen as not to happen	

Allow the children to write events on strips of paper and stick them in the Examples column in the appropriate row.

Interesting mistakes

Probabilities involve subtle use of language. The opposite of 'possible' (may happen) is 'impossible' (cannot happen), but the opposite of 'certain' or 'definite' (100% likely) is 'certain not to happen' or 'definitely won't happen' (100% unlikely).

It is not important to make these subtle distinctions with the children at this stage. They just need to be able to understand possible/impossible and likely/unlikely.

Answers for Pupil Book 3 page 108

<u>Think and share:</u> Individual answers. For example:
Possible: I will read my book.
Impossible: I will travel to Mars.
Might happen: I might play with my friends after school.

1 A definitely won't happen
  B very unlikely to happen
  C probably not
  D equally likely
  E not sure but probably yes
  F quite likely to happen
  G definitely will happen

Possible outcomes

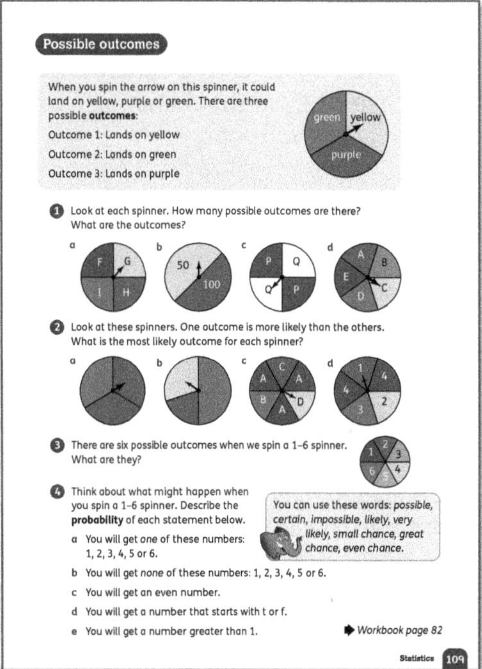

Warm-up

Discuss the spinner at the top of **Pupil Book 3 page 109** with the class. Ask questions such as:
- *What colours could the arrow land on?*
- *Is there more chance of landing on one colour than another? Why or why not?*
- *Could the arrow land on another colour – not green, yellow or purple? Is that possible or impossible?*

Focus

- Discuss the different spinners in question 1 with the children. Let them identify the number of possible *outcomes*.
- The spinners in question 2 are slightly different – on each spinner, one outcome is more likely than the others. The children should notice that if one colour or letter occupies more space on the spinner than others, there is a greater chance of the arrow landing on it.
- Discuss questions 3 and 4 with the class. Let the children work in groups to examine 1–6 spinners and identify the possible outcomes of spinning a 1–6 spinner.

Follow-up

Use **Workbook 3 page 82** as practice and consolidation of the work on possible outcomes.

Challenge

Hand out paper plates or circles cut from card, split pins and marker pens and let the children make their own spinners with two or more equally likely outcomes.

- First, they need to decide how many outcomes they want on their spinner and what the outcomes will look like. They should sketch this on paper before starting to draw or write on the spinner.

- Second, help the children to mark the centre point of the spinner. The children can use rulers to draw lines (diameters and radii, depending on their design).
- The children colour the segments.
- Help the children to attach the cardboard arrow using a split pin. The pin should not be closed too tightly or the arrow will not spin.

Support

If the children are struggling with the probability concepts, stick to simple spinners with just two, three or four equally likely outcomes. It is not essential for the children to work with outcomes that are more or less likely at this stage.

Interesting mistakes

- The children might assume from their work so far that *probability* means working with spinners.
- It is important to keep asking them questions that direct them to realise that probability means the likelihood of different types of events in a range of different situations.
- For each event, ask the children what the possible outcomes are.

Answers for Pupil Book 3 page 109

1 **a** 4 (F, G, H, I)   **b** 2 (50, 100)
   c 2 (P, Q)   **d** 5 (A, B, C, D, E)
2 **a** purple   **b** green   **c** A (red)   **d** 4 (red)
3 1, 2, 3, 4, 5, 6
4 **a** certain   **b** impossible   **c** even chance
   d likely   **e** very likely

Answers for Workbook 3 page 82

1 Individual answers. For example:
   a the sun will rise.   **b** I will eat a snack.
   c I will go to the cinema.
2 Individual answers. For example, a picture of a child flying through the air or picking a £1 coin from a bag that contains only 5p and 10p coins
3 **a** Each section shaded a different colour
   b One section shaded one colour, three sections shaded a different colour
   The children may use patterns or words, numbers or letters instead of colours.

Probability experiments

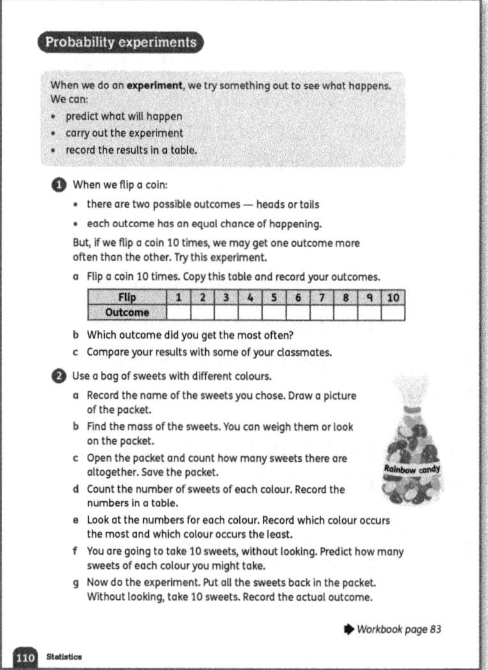

Materials

Coins; packets of sweets with sweets in several different colours (one packet for each group) (alternatively, you could use a 'feely bag' (page 23) of coloured marbles); a weighing scale; colouring pencils

Warm-up

- **Pupil Book 3 page 110** presents two fun, practical experiments. Introduce the term *experiment* and ask the children what they think it means. (An experiment is a practical test, which allows us to get an actual set of outcomes. These outcomes might be different in the next experiment.)
- Explain the simple coin flipping experiment in question 1 and ask the children to predict how many times they think the coin will land on heads and on tails.

Focus

- In question 1, the children flip a coin ten times and record their outcomes in a table.
- They note the outcome that occurs more frequently than the other, and compare their results with other children's results.
- For question 2, if using sweets is inappropriate for your classroom, use a 'feely bag' full of marbles in several different colours. There are two different options for how to present the activity:
  - You may do a single experiment together as the whole class.
  - If you have enough packets of sweets, the children can work in pairs or small groups while you talk them through the instructions.
- Decide on your preferred option, based on how well you feel the children would manage to follow the instructions independently.

- Read through all the instructions with the class. Check what you will need: sweets (or marbles), a weighing scale, colouring pencils.
- The children complete the experiment, and record their answers and results in their notebooks or on **Workbook 3 page 83** as they work through the experiment.
- If they are using their notebooks, they will need to draw a tally table with a row or column for each colour of sweet.

Challenge

Link the sweets (or marbles) experiment to recent work on bar charts. Have the children represent the outcomes of the experiment as a bar chart, with bars representing the numbers of the different colours of sweet.

Some children may like to repeat the sweets experiment at home. They can record the outcomes in the same way as before and draw a bar chart to present their results. They can then compare the two bar charts.

Answers for Pupil Book 3 page 110
1–**2** Individual answers based on practical work

Answers for Workbook 3 page 83
1 Individual answers based on practical work

End-of-unit check

To assess the children's understanding of probability, ask questions such as:
- *When I flip a coin, how many possible outcomes are there?* (2)
- *How many possible outcomes are there when I spin a 1–6 spinner?* (6)
- *What is the probability of landing on an even number with a 1–6 spinner?* (even chance)
- *Can you tell me an event that is very likely/very unlikely/certain to happen today?*
- *Can you tell me an event that is impossible?*
- *Kerry has a bag of 100 marbles. 60 are red, 30 are blue and 10 are yellow. If she takes out a marble without looking, which colour is she least likely to get? Why?* (Yellow because there are fewer yellow marbles.)

UNIT 18 Time

Learning objectives
- Choose the appropriate unit of time for familiar activities.
- Read and record time using digital notation (12-hour clock).
- Interpret and use information in timetables (12-hour clock).
- Understand the difference between a time and a time interval.
- Find intervals between the same units of time in days, weeks, months and years.
- Tell and write time from analogue clocks, including Roman numerals from I to XII, and 12-hour and 24-hour digital clocks.
- Estimate and read time with increasing accuracy to the nearest minute; record and compare time in terms of seconds, minutes and hours; use vocabulary such as o'clock, a.m./p.m., morning, afternoon, noon and midnight.
- Know the number of seconds in a minute and the number of days in each month, year and leap year.
- Compare durations of events (for example, calculate the time taken by particular events or tasks).

Key words

time   o'clock   half past   quarter past/to   analogue clock   digital clock   Roman numerals   second   minute   hour   day   week   month   year   time interval   before   after   earlier   later

Unit introduction

Materials
Large sheet of newspaper or poster paper, coloured marker pens, pictures of different types of clocks and watches

Teaching guidance
Write the word *time* in the middle of a large sheet of newspaper or poster paper. Ask: *What is time?* Let them think about the question and then listen to all their suggestions. Their ideas might include:
- how long things (for example, lessons, games, tasks, events) take
- the time of day
- a number of hours or minutes or seconds.

Encourage the children to think of all the words we use to talk about time. The children might suggest:
- descriptive words such as long, short, ages, quick, slow
- units of time such as *minute, hour, second*

- words such as clock and watch
- words we use when telling the time such as *o'clock, quarter past, half past.*

Create a mind map or poster with all of the children's ideas. As you do so, encourage the children to fill in aspects they may have left out. For example:

- If they have named seconds, hours and minutes, encourage them to think of longer units of time (*days, weeks, months, years*).
- Ask them to think of what we use to measure and record time (clocks, stopwatches, calendars and timetables).

The children can decorate the poster with pictures of clocks, watches and calendars, and you can display the finished poster on the classroom wall.

Reading time

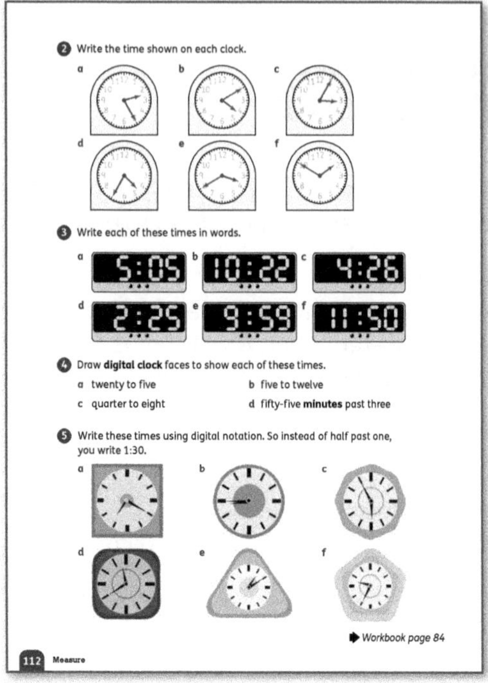

Materials

Clocks with moveable hands (these can be easily made using paper plates, hands cut from card and split pins or paper fasteners); laminated blank clock faces (analogue and digital), whiteboard markers, cloths or paper towels for wiping the clock faces

Warm-up

Start with one or two practical activities:

- Working in pairs, the children take turns to say a time (using *o'clock, half past, quarter past* or *quarter to*). The other child moves the hands on a clock face to show the time.
  They can then take turns to make a time on the clock face for the other child to say.
  You may need to demonstrate a few examples first to refresh the children's memory about how we read times to the hour, half-hour and quarter-hour.
- Working in pairs, the children use laminated blank digital clock faces. One child says the time and the other writes it on the clock using a whiteboard marker. They can then use a cloth to wipe it off, and swap roles.

Focus

- Point out to the children that the hour hand on an analogue clock goes around twice during the course of a day: once between midnight and midday and once between midday and midnight. Demonstrate this on a clock face with moveable hands. Ask: *What problem do we have when we tell the time like this?*
- The children should notice that each time is repeated, so we need a way of telling the difference between times in the morning and times in the afternoon or evening. Introduce the terms *a.m.* and *p.m.*
- Say some different times and let the children say whether each time is in the morning or afternoon/evening. Ask: *Is (2 a.m./4 p.m./6 a.m. etc.) an afternoon time or a morning time?*
- Show a time on an *analogue* or *digital clock* face and say whether it is morning or afternoon/evening. Ask the children to say the time using a.m. or p.m.
- Point out that the face of an analogue clock is divided into 12. Because there are 60 minutes in an hour, each division represents 5 minutes. Show the positions of the hands of a clock at 5 past the hour, 10 past the hour, quarter past the hour, 20 past the hour, 25 past the hour and half past the hour. When you are certain that the children have grasped these concepts, show 25 to the hour, 20 to the hour, quarter to the hour, 10 to the hour and 5 to the hour.
- <u>Think and share:</u> Discuss the picture and questions at the top of **Pupil Book 3 page 111**.
- All the clocks show 1 o'clock, but some of them indicate that it is the afternoon or morning.
- If any of the watch or clock formats shown are unfamiliar to the children, show them some examples and explain how each format works.

- Let the children work through questions 1–5 on
  Pupil Book 3 page 111 and **page 112**, independently
  if possible. Assist them as needed.

Follow-up
Use **Workbook 3 page 84** to provide additional practice
and consolidation.

Challenge
Give the children problems such as these to solve:
- *The time is now half past 4. Where will the hour hand
  be pointing in 1 and a half hours? (6)*
- *The time is now quarter past 7. Where will the minute
  hand be pointing in 20 minutes? (25 to or, possibly, to
  the 7)*

Support
Some children may be used to digital watches, but not
analogue clocks. Give them plenty of practice in making
given times on analogue clock faces with moveable
hands.

It is also useful to have a wall clock in the classroom.
Help the children to note the time at various times of
the day. You can ask question such as:
- *How much longer until the clock is on the hour again?*
- *How many minutes past the hour is it now?*
- *What will the time be in 10/15/25 minutes?*

Interesting mistakes
Statements such as '11:50 is 10 to 11' indicate a very
common error because we tend to read the hour first.

Work through the error by referring to a more familiar
time first: *What time is 11:30? Is it before or after
11 o'clock? OK, so the minutes after the dots tell us how
long has passed after the hour. How long has passed
when it is 11:50?*

It is useful to give examples of other, similar mistakes
for the children to correct using their own reasoning.
For example, say: *The following statements are incorrect.
Can you explain why?*
- '06:45 is quarter past 6' (It is quarter to 7. They have
  read the hours, i.e. 6 first but it 45 minutes past 6
  which is the same as 15 minutes or a quarter to 7.)
- '12:05 is 12 minutes past 5'. (They have got the hours
  and minutes mixed up.)

Continue with similar examples.

Answers for Pupil Book 3 pages 111–112
Think and share: The clocks that show exactly the
same time of day are the digital clocks and watches
that show 13:00 and 01:00 p.m.

The clocks that *could* show different times of day are
both the analogue clocks and the digital clock that
just says 01:00. We do not know if it is a.m. or p.m.

(The digital watch that shows 01:00 a.m. *definitely*
shows a different time of day from those that show
13:00 and 01:00 p.m.)

One hour after the time shown: 02:00 p.m. or 14:00

Individual answers to the notice and wonder
questions.

1   a   b   b   half past three
  c   Individual drawings of clocks showing
7 o'clock

2   a   twenty-five past two
  b   ten past four
  c   five past three   d   twenty-five to five
  e   twenty to four   f   ten to two

3   a   five past five
  b   twenty-two minutes past ten
  c   twenty-six minutes past four
  d   twenty-five past two
  e   one minute to ten   f   ten to twelve

4   Drawings of digital clocks with these times:
  a   04:40   b   11:55   c   07:45   d   03:55

5   a   7:20   b   8:45   c   5:55
  d   11:40   e   1:10   f   9:35

Answers for Workbook 3 page 84

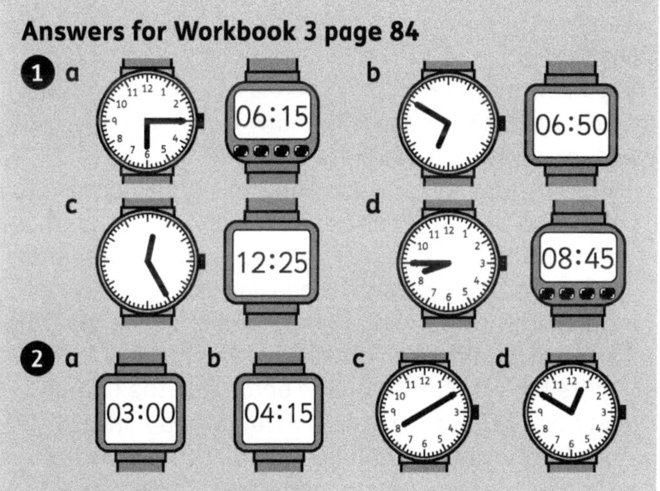

Analogue clocks

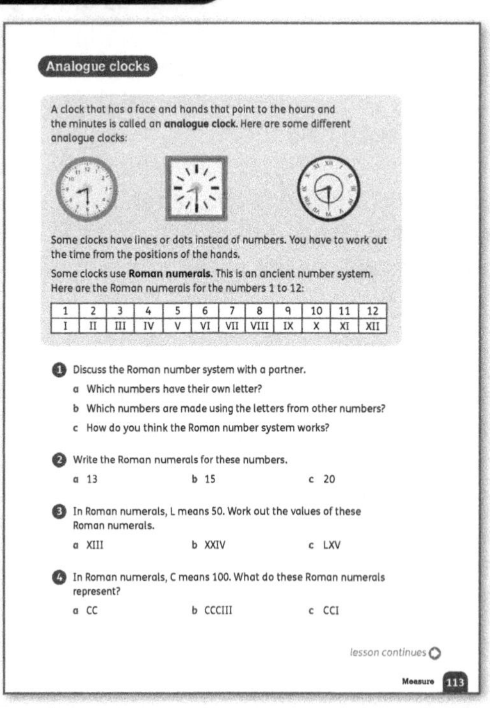

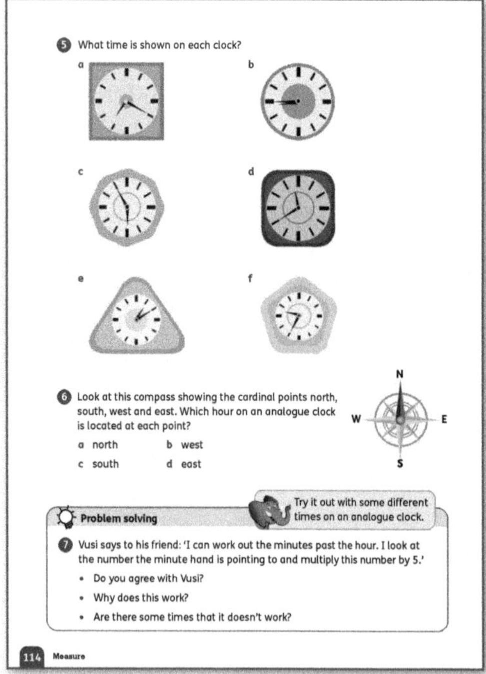

Materials

Clock with Roman numerals, with moveable hands

Warm-up

You can introduce *Roman numerals* as a code-cracking activity, as this is a code with many clues. Here are some ideas for working in this way:

- Show a clock with Roman numerals to the class. Ask the children if they have seen these kinds of numbers before. They may be able to guess what the numbers mean from their positions on the clock face. You could start by pointing to the number at the top (XII) and asking: *Who can guess what this number is?* (12)
- Write the number 12 on the board and the Roman numeral XII next to it. Ask: *What do you notice?* If necessary, give the children the hint that I means 1 in this number system. Ask: *How do you think we show the number 2?* (II)
- *What do you think X means?* Guide the children to work out that the number is made up of the ten, represented by X, plus 2 ones.
- Next, let the children guess the Roman numeral for 11 (XI). They can use the position of the number on the clock face to help them.
- Next, show IX. Say: *This is 9. How do you think this number works?* (9 is 'one before ten'.)
- Let the children work out the rest of the Roman numerals, as well as the pattern for writing the numbers.

Focus

- Work through the information at the top of **Pupil Book 3 page 113** with the class.
- Working in pairs, the children can answer question 1 orally.
- The children can work through questions 2–7 on **Pupil Book 3 page 113** and **page 114** independently or in pairs.

- <u>Problem solving:</u> For question 7, the children could try out different times on a clock with moveable hands to see whether Vusi is correct.

Challenge

Some children may enjoy finding out how to write the Roman numerals for larger numbers, up to 100 or 200.

Support

Give the children cards with the first 12 Roman numerals in mixed-up order. They can reorder them correctly, discussing how they worked out the order.

Interesting mistakes

- Ask the children to identify the mistake in 'VIII is greater than XII because it has more digits'.
- Encourage them to notice that, unlike our number system, the Roman system is not a place-value system.
- Draw the children's attention to how many digits there are in each of the Roman numerals 1 to 5. Let them notice how the number of digits increases and then decreases again.

Answers for Pupil Book 3 pages 113–114

1 a 1, 5, 10

   b 2, 3, 4, 6, 7, 8, 9, 11, 12

   c Possible answers: The Roman number system uses patterns or groups of the symbols for one, five and ten.
   There can be 1, 2 or 3 ones together. There can be up to 3 ones after a five or ten, or 1 one before it.
   A one before a five or ten is subtracted from the five or ten.
   Ones after a five or ten are added to the five or ten.

2 a XIII   b XV   c XX

3 a 13   b 24   c 65

4 a 200   b 303   c 164

5 a eleven o'clock   b quarter past twelve
   c half past two   d twenty past ten
   e twenty-five to four   f ten to one

6 a 12 o'clock   b 9 o'clock
   c 6 o'clock   d 3 o'clock

7 Yes.
Possible answer: There are 60 minutes in an hour and 12 hours on the clockface. 60 ÷ 12 = 5, so each hour interval represents 5 minutes.
It does not work for 12, as this is 0 minutes past the hour, not 60.

Digital times

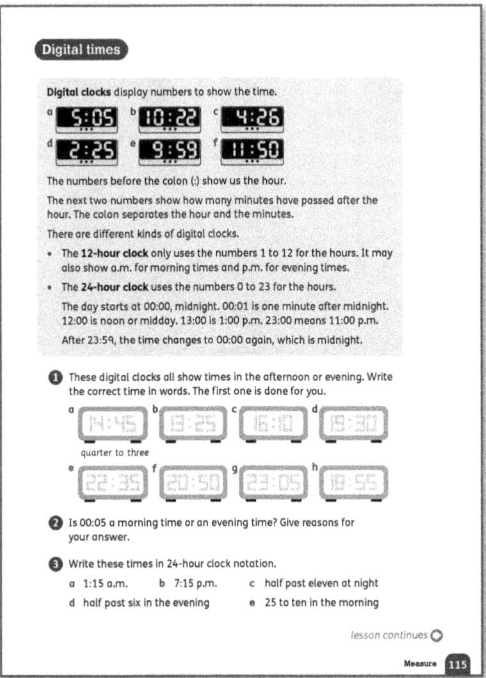

Digital times

Digital clocks display numbers to show the time.

a 5:05 b 10:22 c 4:26
d 2:25 e 9:59 f 11:50

The numbers before the colon (:) show us the hour.

The next two numbers show how many minutes have passed after the hour. The colon separates the hour and the minutes.

There are different kinds of digital clocks.

- The **12-hour clock** only uses the numbers 1 to 12 for the hours. It may also show a.m. for morning times and p.m. for evening times.
- The **24-hour clock** uses the numbers 0 to 23 for the hours.

  The day starts at 00:00, midnight. 00:01 is one minute after midnight. 12:00 is noon or midday. 13:00 is 1:00 p.m. 23:00 means 11:00 p.m.

  After 23:59, the time changes to 00:00 again, which is midnight.

1 These digital clocks all show times in the afternoon or evening. Write the correct time in words. The first one is done for you.

a 14:45 b 13:25 c 16:10 d 19:30

quarter to three

e 22:35 f 20:50 g 23:05 h 18:55

2 Is 00:05 a morning time or an evening time? Give reasons for your answer.

3 Write these times in 24-hour clock notation.

a 1:15 a.m. b 7:15 p.m. c half past eleven at night
d half past six in the evening e 25 to ten in the morning

lesson continues

Measure **115**

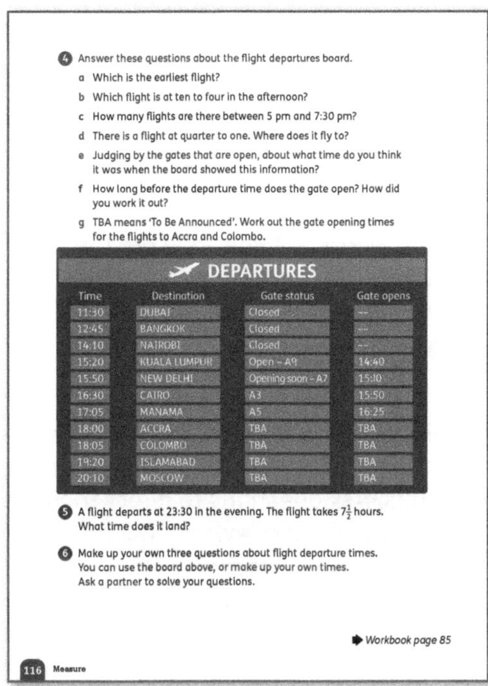

4 Answer these questions about the flight departures board.

a Which is the earliest flight?

b Which flight is at ten to four in the afternoon?

c How many flights are there between 5 pm and 7:30 pm?

d There is a flight at quarter to one. Where does it fly to?

e Judging by the gates that are open, about what time do you think it was when the board showed this information?

f How long before the departure time does the gate open? How did you work it out?

g TBA means 'To Be Announced'. Work out the gate opening times for the flights to Accra and Colombo.

✈ DEPARTURES

Time	Destination	Gate status	Gate opens
11:30	DUBAI	Closed	--
12:45	BANGKOK	Closed	--
14:10	NAIROBI	Closed	--
15:20	KUALA LUMPUR	Open – A9	14:40
15:50	NEW DELHI	Opening soon – A7	15:10
16:30	CAIRO	A3	15:50
17:05	MANAMA	A5	16:25
18:00	ACCRA	TBA	TBA
18:05	COLOMBO	TBA	TBA
19:20	ISLAMABAD	TBA	TBA
20:10	MOSCOW	TBA	TBA

5 A flight departs at 23:30 in the evening. The flight takes $7\frac{1}{2}$ hours. What time does it land?

6 Make up your own three questions about flight departure times. You can use the board above, or make up your own times. Ask a partner to solve your questions.

➡ *Workbook page 85*

116 Measure

Materials

Laminated blank digital clock faces, whiteboard markers, cloths or paper towels for wiping the clock faces (alternatively, you can use an easily adjusted large digital clock)

Warm-up

- The children have already worked with digital clock times. In this lesson, they will begin to work with 24-hour clock times.
- Discuss the information at the top of **Pupil Book 3 page 115** with the class.

Focus

- Use laminated blank digital clock faces to show some different times.
- Explain how 24-hour clocks work: *After 12 noon, we use the numbers 13 to 23 to show the hours from 1 in the afternoon to 11 at night.*
- Demonstrate some times and then follow up with questions. For example, change the time on the digital clock face to 13:00 and explain that this is 1 in the afternoon. You could ask: *Who wants to show 5 minutes past 1 in the afternoon on this clock? After another 10 minutes, what time will it be? Write this time for me.*
- Next show 13:30, then 13:35. Remind the children that 35 minutes past the hour is the same as 25 to the next hour. Ask: *What will be the next hour after 1 o'clock? (2 o'clock) How do we write 2 o'clock in the afternoon using 24-hour time? If 1 o'clock is 13:00, 2 o'clock will be … ?* (14:00), and so on.
- Work through the times, and spend some time on how we write midnight (00:00). Work through the times between midnight and 01:00 with the class.
- Once the children have had plenty of practice with 24-hour clock times, let them work through questions 1–6 on **Pupil Book 3 page 115** and **page 116**. Question 4 introduces work with timetables.

Follow-up

The children can practise 12-hour digital clock times and work on a timetable in more depth using **Workbook 3 page 85**.

Challenge

Have the children work with real timetables such as ferry timetables or TV guides. You can ask questions about these.

Support

- Some children may need extra help to become familiar with the 24-hour clock, particularly the times from 13:00 to 23:00.
- Practise reading and showing on-the-hour times before moving on to half-past-the-hour times.
- Once the children are more familiar with 24-hour notation for these times, they will gradually become more comfortable with times to the minute.

Interesting mistakes

- Give the children this mistake: '*13:05 is 5 past 3.*' Have the children explain the mistake and how to correct it.
- This is a very common error when working with 24-hour times and one that many adults also make – for example mistaking 17:00 for 7 o'clock or 19:00 for 9 o'clock.
- Practise counting on the hours from 12 o'clock. It may also help the children to notice that 12 + 1 = 13, 12 + 2 = 14, 12 + 3 = 15, and so on. These number bonds can help them to recall the 24-hour clock times correctly.

Answers for Pupil Book 3 pages 115–116

1 **a** quarter to three (provided as an example)
 b twenty-five past one
 c ten past four
 d half past seven
 e twenty-five to eleven
 f ten to nine
 g five past eleven
 h five to seven

2 Possible answer: It is a morning time, as it is five past midnight, so very early in the morning.

3 **a** 01:15 **b** 19:15 **c** 23:30
 d 18:30 **e** 09:35

4 **a** Dubai **b** New Delhi
 c 4 **d** Bangkok
 e Any time between 14:40 and 15:10
 f 40 minutes The time between the times in the 'Gate opens' and 'Time' columns for each flight.
 g Accra 17:20 Colombo 17:25

5 07:00

6 Individual answers

Answers for Workbook 3 page 85

1 **a** 5:00 p.m. **b** 3:30 a.m.
 c 6:45 a.m. **d** 3:40 p.m.

2 **a** twenty-five to seven in the evening
 b half past twelve in the afternoon
 c quarter past eleven in the morning

3 **a** 5 minutes **b** 10 minutes
 c 15 minutes **d** 25 minutes

4 every half hour

5 **a** Bus C **b** 11:40 a.m.

Units of time

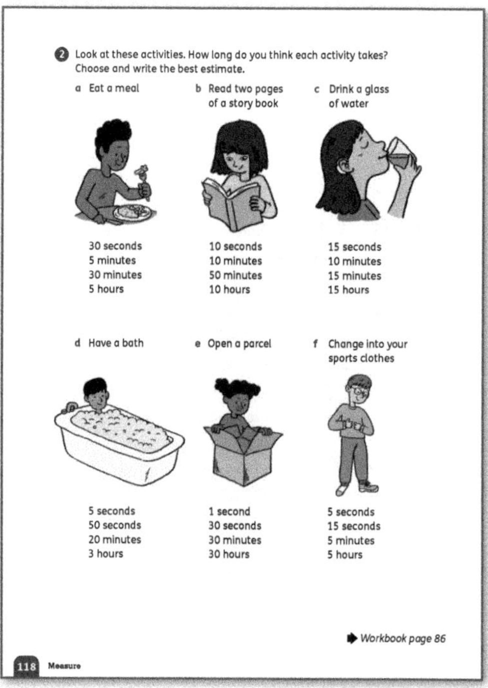

Warm-up

Discuss how long different activities take. Ask questions such as:

- *How long do you usually take to eat breakfast in the morning?*
- *How long are our maths lessons?* (If necessary, note the start and end time, and help the children to work out the duration.)
- *How long might a party last?*
- *How long is a weekend? How long are the school holidays? What about mid-term break?*
- *How long do you usually spend at the park/ beach/playground?*

Focus

- Let the children discuss activities that take different amounts of time and that we measure using different units. Ask questions to direct them, such as:
  - What can you do in a few seconds?
  - What takes a whole year?
  - How long is it until the end of Year 3?
- Turn to **Pupil Book 3 page 117**. Discuss the examples shown at the top of the page. Revise the different units of time. If necessary, remind the children of the number of days in each month and the difference between a year and a leap year.
- Once the children have spent some time talking about the durations of a variety of activities and the different units of time, they can work through the estimating and timing activities in questions 1 and 2 on **Pupil Book 3 page 117** and **page 118**.

Follow-up

The children can work through further estimating and timing activities on **Workbook 3 page 86**.

Answers for Pupil Book 3 pages 117–118

1 Individual answers

2
a 30 minutes	**b** 10 minutes
c 15 seconds	**d** 20 minutes
e 30 seconds	**f** 5 minutes

Answers for Workbook 3 page 86

1 Individual answers

Time and time intervals

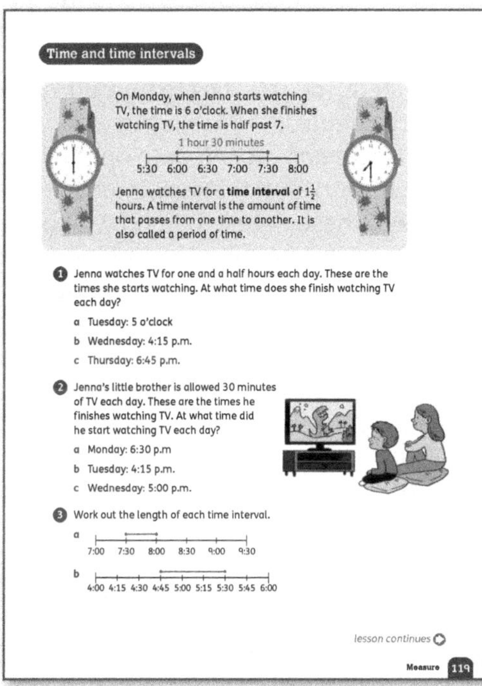

Time and time intervals

On Monday, when Jenna starts watching TV, the time is 6 o'clock. When she finishes watching TV, the time is half past 7.

1 hour 30 minutes

5:30   6:00   6:30   7:00   7:30   8:00

Jenna watches TV for a **time interval** of $1\frac{1}{2}$ hours. A time interval is the amount of time that passes from one time to another. It is also called a period of time.

1 Jenna watches TV for one and a half hours each day. These are the times she starts watching. At what time does she finish watching TV each day?

a   Tuesday: 5 o'clock

b   Wednesday: 4:15 p.m.

c   Thursday: 6:45 p.m.

2 Jenna's little brother is allowed 30 minutes of TV each day. These are the times he finishes watching TV. At what time did he start watching TV each day?

a   Monday: 6:30 p.m

b   Tuesday: 4:15 p.m.

c   Wednesday: 5:00 p.m.

3 Work out the length of each time interval.

a

7:00   7:30   8:00   8:30   9:00   9:30

b

4:00   4:15   4:30   4:45   5:00   5:15   5:30   5:45   6:00

lesson continues

Measure **119**

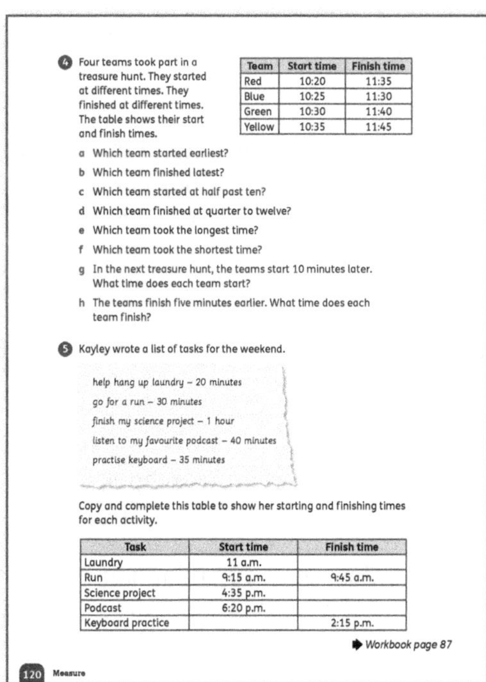

4 Four teams took part in a treasure hunt. They started at different times. They finished at different times. The table shows their start and finish times.

Team	Start time	Finish time
Red	10:20	11:35
Blue	10:25	11:30
Green	10:30	11:40
Yellow	10:35	11:45

a   Which team started earliest?

b   Which team finished latest?

c   Which team started at half past ten?

d   Which team finished at quarter to twelve?

e   Which team took the longest time?

f   Which team took the shortest time?

g   In the next treasure hunt, the teams start 10 minutes later. What time does each team start?

h   The teams finish five minutes earlier. What time does each team finish?

5 Kayley wrote a list of tasks for the weekend.

help hang up laundry – 20 minutes

go for a run – 30 minutes

finish my science project – 1 hour

listen to my favourite podcast – 40 minutes

practise keyboard – 35 minutes

Copy and complete this table to show her starting and finishing times for each activity.

Task	Start time	Finish time
Laundry	11 a.m.	
Run	9:15 a.m.	9:45 a.m.
Science project	4:35 p.m.	
Podcast	6:20 p.m.	
Keyboard practice		2:15 p.m.

➡ *Workbook page 87*

120 Measure

Materials
Clock with moveable hands

Warm-up
Show some different times on a clock with moveable hands and ask the children: *What time is it?*

Focus
- Turn to **Pupil Book 3 page 119** and introduce the term *time interval,* which is the amount of time between a start time and a finish time.
- Work through the example at the top of this page with the class. Show the start time on a clock with moveable hands. Demonstrate how the hands of the clock move to get from 6 o'clock to half past 7. Break this down into:
  - one whole hour from 6 o'clock to 7 o'clock (move the long hand all the way around the clock and move the small hand to the next hour)
  - half an hour from 7 o'clock to half past 7 (move the long hand halfway around the clock and move the small hand to halfway between 7 and 8).
- You can work through some of questions 1–5 on **Pupil Book 3 page 119** and **page 120** together with the class, then let them continue independently.

Follow-up
- Use **Workbook 3 page 87** to consolidate the work covered in this lesson.

It is important that the children can work flexibly with time intervals. If the children need additional practice, you can:
- Give a start time and a time interval – the children work out the finish time.
- Give a start time and a finish time – the children work out the time interval.
- Give a time interval and a finish time – the children work out the start time.
- Give one start time and some different finish times – the children work out who finished first, second, last, etc.
- Give a list of different events with different start times and different time intervals – the children identify which event finishes the earliest/latest.

Support
- Spend more time working with a clock with moveable hands to help the children work out the time interval between two times.
- Start with whole-hour times and then half-hour times.
- Then give the children extra practice in calculating times to 5 minutes. For example, ask: *How many minutes is it from 10:25 to 10:50/8:55 to 9:05/4:15 to 4:45/1:20 to 2:10?* (25/10/30/50)

Interesting mistakes
Invite the children to correct what is wrong in this statement: *'1:20 to 2:10 takes 10 minutes, because the difference between 20 and 10 is 10.'* Draw their attention to the need to take both the minutes and the hours into account.

Answers for Pupil Book 3 pages 119–120

1 a 6:30 p.m. b 5:45 p.m. c 8:15 p.m.

2 a 6 p.m. b 3:45 p.m. c 4:30 p.m.

3 a 30 minutes or half an hour
b 45 minutes or three-quarters of an hour

4 a Red b Yellow c Green
d Yellow e Red f Blue
g Red 10:30, Blue 10:35, Green 10:40, Yellow 10:45
h Red 11:30, Blue 11:25, Green 11:35, Yellow 11:40

5 Laundry finish time: 11:20 a.m.
Science project finish time: 5:35 p.m.
Podcast finish time: 7 p.m.
Keyboard practice start time: 1:40 p.m.

Answers for Workbook 3 page 87

1 Sunan: pointer to 30   Ayesha: pointer to 45
Taha: pointer to 35   Kehinde: pointer to 40
Nilar: pointer to 25   Rubina: pointer to 50

2 50 minutes, 45 minutes, 40 minutes, 35 minutes, 30 minutes, 25 minutes

3 25 minutes

4 Ayesha

5 Nilar

Calendars and dates

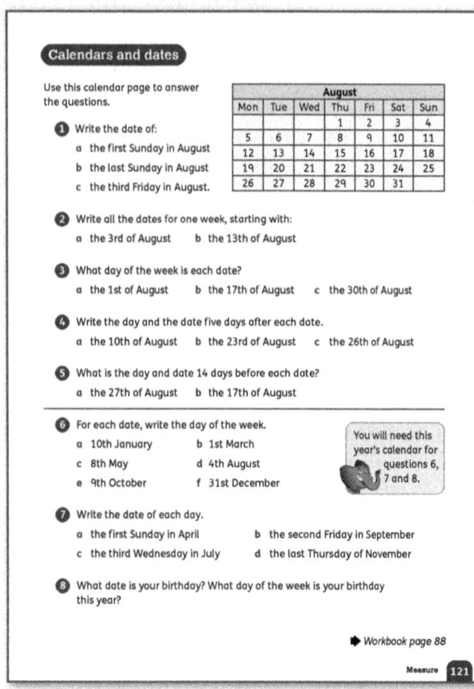

Materials
A calendar

There are different formats for writing dates. Decide on the format you want the children to use, based on what is usual in your country. Some different formats include: 4th August 2022, 4 August 2022 or 04/08/22.

Most countries give dates in the order day, month year. However, in the USA, dates are given in the order month, day, year, so the same date would be 08/04/22 or August 8, 2022.

Warm-up
Here are some ideas for activities to introduce working with dates and calendars:
- Write a start date and an end date on the board. Ask the children to work out: first how many days there are in between these dates, then how many school days. Allow them to consult a calendar if necessary and encourage them to share their answers. Discuss how you can check this and do so.
- Ask the children to work out dates that involve counting in tens. Ask the children questions such as: *It is the 2nd of June. What is the date 10 days later?* (12th of June)
Avoid bridging months unless you allow the children to work with a calendar.
- Ask questions based on the current date. You can have a calendar on display to help with this. For example, say: *Starting this week, write:*
  o *the dates of the next four Fridays*
  o *the date on Saturday*
  o *the dates of the next three Tuesdays*
  o *the date a week ago today.*

Focus
- Make sure the children understand that a date is a specific day of a month. They also need to understand the terms *before*, *after*, *earlier* and *later*, in phrases such as 'one day after', 'one day before', 'one week later' and 'one week earlier'.
- Once the children have grasped these concepts, they can work independently through questions 1–8 on **Pupil Book 3 page 121**.

Follow-up
There are further questions on **Workbook 3 page 88**.

Challenge
Give the children this year's calendar and ask them to work out on which day of the week today's date was:
- 2 years ago
- 5 years ago
- 10 years ago.

They will need to take into account the number of days in a year (including any leap years). Challenge the children to find a pattern for working out the day of a particular date in a past or future year.

You can also give the children maths riddles. Start with easier ones, such as:

- *In three days' time, it will be Sunday. What day is it today?*
- *Yesterday was two days before Tuesday. What day is it today?*

Move on to trickier ones, such as:

- *Three days ago, yesterday was the day before Wednesday. What day will it be tomorrow?*

Answers for Pupil Book 3 page 121

1 a 4th August　　**b** 25th August
　c 16th August
2 a 3rd August, 4th August, 5th August, 6th August, 7th August, 8th August, 9th August
　b 13th August, 14th August, 15th August, 16th August, 17th August, 18th August, 19th August
3 a Thursday　　**b** Saturday　　**c** Friday
4 a Thursday, 15th August
　b Wednesday, 28th August
　c Saturday 31st August
5 a Friday, 23rd August　　**b** Saturday, 3rd August
6–8 Individual answers, using the dates for the current year

Answers for Workbook 3 page 88

1 **February**

Mon	Tues	Wed	Thur	Fri	Sat	Sun
		1	2	3	4	5
6	7	8	9	10	11	12
13	14	15	16	17	18	19
20	21	22	23	24	25	26
27	28	(29)				

September					
Sun	1	8	15	22	29
Mon	2	9	16	23	30
Tues	3	10	17	24	
Wed	4	11	18	25	
Thurs	5	12	19	26	
Fri	6	13	20	27	
Sat	7	14	21	28	

2 Individual answers

Calendar problems

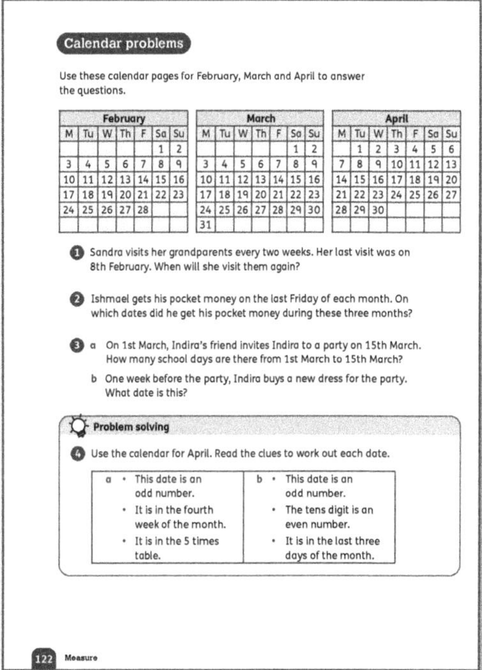

Warm-up

Ask questions such as:

- *Does a calendar show minutes and hours?* (no) *What units of time does a calendar show?* (days, weeks and months)
- *How do we use a calendar?*
- *How many months are there in a year?* (12)
- *What do the numbers on a calendar tell us?* (the dates, specific days in a month)
- *How many days make a week?* (7)
- *How many weeks does this month have?*
- *Which week will continue into the next month?*

Focus

- **Pupil Book 3 page 122** provides more practice in using calendars. The children can work independently on questions 1–3.
- <u>Problem solving:</u> For question 4, the children can work in pairs to work out the mystery dates.

Answers for Pupil Book 3 page 122

1 22nd February
2 28th February, 28th March, 25th April
3 a 10 days　　**b** 8th March
4 a 25th April　　**b** 29th April

To assess the children's understanding of time, ask questions such as:

- *What time is it when the long hand points to 12 and the short hand points to 3? (3 o'clock)*
- *What time is it when the clock shows 10:30? (half past 10)*
- *When the time is 04:15, is it quarter past 4 or quarter to 4? (quarter past 4)*
- *Does 8 a.m. mean 8 o'clock in the morning or 8 o'clock in the evening? (8 o'clock in the morning)*
- *What time is it when the short hand points to just before the 2 and the long hand points to 10? (ten to 2)*
- *Is 08:35 the same as 25 past 8 or 25 to 9? (25 to 9)*
- *School starts at 8:15. Lunch is at 12 o'clock. How many hours and minutes are there between the start of school and lunchtime? (3 hours and 45 minutes)*
- *Which month comes between June and August? (July)*
- *How many days is it from (date) to (date)?*
- *How many weeks is it until the school holidays start?*

Give the children this time quiz to test their knowledge of time units and vocabulary:

- *How many weeks are there in a year? (52)*
- *How many days are there in two weeks? (14)*
- *How many months are there in two years? (24)*
- *What is the sixth month of the year? (June)*
- *How many days are there in a year? (365)*
- *How is a leap year different from other years? (It has an extra day in February so it has 366 days.)*
- *Which month comes before September? (August)*
- *Which day is two days after Thursday? (Saturday)*
- *What is the first day of the week? (The answer will depend on your country.)*
- *How many weeks are there in half a year? (26)*
- *How many weeks are there in quarter of a year? (13)*
- *Which is the shortest month? (February)*

Ask the children to write or show these times:

- *15 minutes before 2 o'clock (quarter to 2 or 1:45)*
- *5 minutes after 1:15 (twenty past 1 or 1:20)*
- *10 minutes before 5:30 (twenty past 5 or 5:20)*
- *half an hour after 4 o'clock (half past four or 4:30).*

Mixed practice 3

You can use Mixed practice 3 on **Pupil Book 3 pages 123–124** to assess the children's confidence in the concepts from Units 13–18.

Answers for Mixed practice 3, Pupil Book 3 pages 123–124

1 A straw

2 a True   b False. A cube has 6 faces.

3 a East   b South   c North

4 D

5 a $\frac{1}{4}$ b $\frac{1}{3}$ c $\frac{1}{5}$ d $\frac{1}{6}$

6 a The two sections are not the same size.
    b The four sections are not the same size.
    c The arrow is showing 2 intervals out of a total of 6 intervals, so is showing $\frac{2}{6}$.

7 a $8   b $\frac{3}{4}$

8 a 5   b 5   c 6

9 $\frac{2}{8}, \frac{3}{8}, \frac{5}{8}, \frac{6}{8}, \frac{7}{8}$

10 0.75 m

11 a $\frac{3}{5}$ b $\frac{3}{10}$ c $\frac{5}{10}$

12 Salvina has added both the numerators and the denominators. She should have only added the numerators and left the denominator the same.

13 200 ml, 500 ml, $\frac{3}{4}$ litre, 900 ml, 1 litre

14 200 ml

15 25 °C

16 70 °C

17 Drawing of a circle divided into four equal sectors that are all different (for example, shaded different colours or with different patterns, numbers or letters)